# THE SUBMARINE

## AN ILLUSTRATED HISTORY FROM 1900 TO 1950

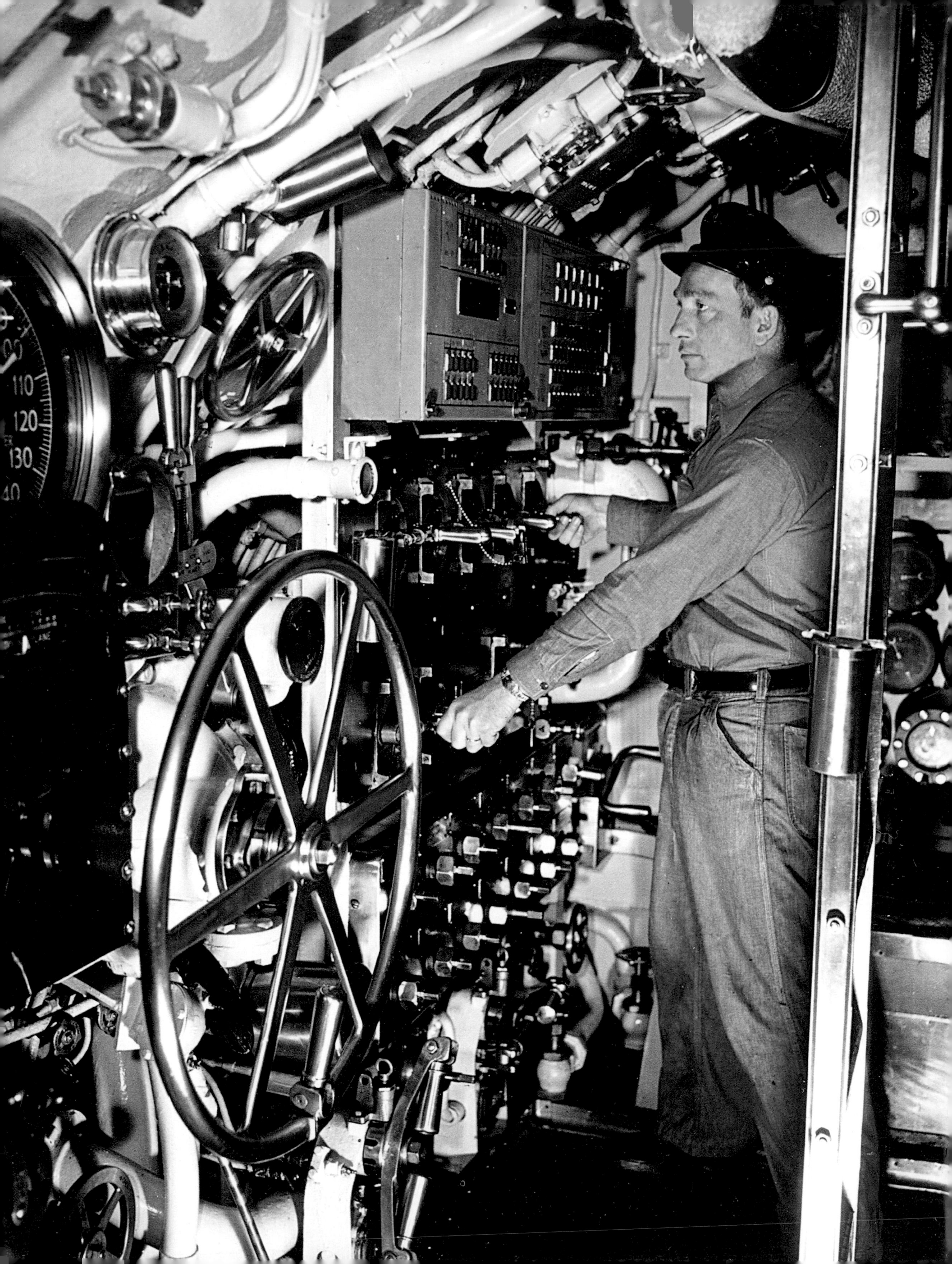
110
120
130

# THE SUBMARINE

## AN ILLUSTRATED HISTORY FROM 1900 TO 1950

AN AUTHORITATIVE GUIDE TO THE DEVELOPMENT OF UNDERWATER VESSELS AROUND THE WORLD, WITH OVER 400 HISTORICAL PHOTOGRAPHS, PAINTINGS AND CUTAWAYS

JOHN PARKER

southwater

This edition is published by Southwater
an imprint of Anness Publishing Ltd
Hermes House, 88–89 Blackfriars Road, London SE1 8HA
tel. 020 7401 2077; fax 020 7633 9499

www.southwaterbooks.com; www.annesspublishing.com

Anness Publishing has a new picture agency outlet for images for publishing, promotions or advertising. Please visit our website www.practicalpictures.com for more information.

UK agent: The Manning Partnership Ltd
tel. 01225 478444; fax 01225 478440; sales@manning-partnership.co.uk
UK distributor: Grantham Book Services Ltd
tel. 01476 541080; fax 01476 541061; orders@gbs.tbs-ltd.co.uk
North American agent/distributor: National Book Network
tel. 301 459 3366; fax 301 429 5746; www.nbnbooks.com
Australian agent/distributor: Pan Macmillan Australia
tel. 1300 135 113; fax 1300 135 103; customer.service@macmillan.com.au
New Zealand agent/distributor: David Bateman Ltd
tel. (09) 415 7664; fax (09) 415 8892

Publisher: Joanna Lorenz; Senior Editor: Felicity Forster
Copy Editor and Indexer: Tim Ellerby
Designer: Design Principals
Production Controller: Mai Ling Collyer



1 3 5 7 9 10 8 6 4 2

A CIP catalogue record for this book is available from the British Library.

Previously published as part of a larger volume,
*The World Encyclopedia of Submarines.*

**NOTE:** The nationality of each submarine is identified in the specification box by the national flag that was in use at the time of the vessel's commissioning and service.

# Contents

TOP AND ABOVE: **The progression of submarine development is perhaps no better demonstrated by the size comparison of the world's largest submarine, the Russian Typhoon class (TOP), crewed by 160 men and having a submerged weight of 33,787 tonnes/33,253 tons, and the first American *Holland* (ABOVE), which was a mere 72.83 tonnes/71.7 tons submerged and was crowded with a full crew of nine.**

# Introduction

As military forces across the world geared up for the incredible succession of truly devastating developments that emerged through the 20th century, each one scaling up the power of destruction to previously unimagined levels, two key elements were prominent: invention and courage of an outstanding order. Among those groups handsomely endowed with both these qualities, the Submarine Service must surely rank highly. Invention and courage came in abundance, especially in the early years when the two words were linked by the very deed. Then, submariners were considered reckless fools, to put it mildly, for simply agreeing to going out in such death-traps. When the fighting of war was added to the equation, volunteers were considered to be quite mad, especially as gruesome stories of lost submarines in peacetime began to emerge. But, when all was said and done, the future of humankind literally lay in their hands.

This is no overstatement. From virtually a standing start at the beginning of the 20th century, submarines were rapidly developed through the patience and ingenuity of a handful of dedicated engineers and designers who were often financially challenged, receiving little support from naval chiefs and governments. Within just a couple decades of rapid progress, the so-called Silent Service became the new Navy and Masters of the Sea, culminating in an exceedingly volatile head-to-head nuclear armament. The extent of that progression is perhaps best illustrated by the size factor. In 1900, America's first submarine, the *Holland*, had a submerged weight of 76 tonnes/75 tons, carried two tiny torpedoes and a crew of nine, whereas the Russian Typhoon SSBN of the new era of submarines weighed in at 33,000 tonnes/32,479 tons, was crewed by 160 and carried 20 nation-obliterating ballistic missiles. By the middle of the 20th century, the giant boats of the opposing forces, collectively carrying more explosive power than was unleashed in both World Wars up to and including Hiroshima, were operating in close proximity along the flashpoint sea lanes of the Cold War that emerged in the wake of World War II.

The years of development that brought submarines to the very forefront of world conflagration are recorded in detail, a 50-year time span that saw incredible advances in submarine warfare that were to pave the way for the new post-war era of multi-billion-dollar boats. These carry the mission forward into the next level of submarine capability, now to be coupled with a more diverse range of firepower to provide an effective and flexible response to the new emerging threats to national interests and the general population.

**The influence of early experimental submarine technology was undoubtedly evident in the dramatic developments in post-1950's design and capability, demonstrated here with a "before and after" line-up of boats. The awesome power that was eventually achieved can be seen from the open tubes of USS *Sam Rayburn* (LEFT), one of the new breed of Polaris missile-carrying boats, and conversely the interesting Japanese non-nuclear boats such as the 1975 Yushio class of the teardrop design (ABOVE) which emanated from wartime experimental craft. But the idea of submarines to replace battleships dated to the 1920s with the hugely attractive British boat, *M2* (BELOW LEFT), which was the first to carry substantial armoury in the shape of a 305mm/12in gun ahead of a conning tower which itself was based on that of the majestic old dreadnought battleships. But, of course, it was the sleek German *XX1* (BELOW), built towards the end of World War II, that provided the Allies with a veritable wealth of technology.**

Within these pages, it will be seen that decades of invention in the first half of the 20th century paved the way, at great human cost, for the new breed of submarines developed in the second half of that century. Those early decades saw technological innovations that ultimately led to the arrival of the true submarine, envisioned by Jules Verne in his novel *Twenty Thousand Leagues under the Sea*, picked up and explored by heroic and dedicated inventors, technicians and crews who continued the mission.

The journey towards unlimited submersibility through the introduction of nuclear propulsion has been just one part – albeit the most vital – of the multi-layered and international saga of the submarine, often spurred by war and the quest for superiority, but generally advanced by building upon the dreams, ideas and inventions of a relatively small number of men of various nationalities.

The range and diversity of boats that emerged during the first half-century of submarine development stands as a historical tribute to the inventors, designers, builders and, not least, those who served on them. Furthermore, the motives and ambitions of the nations that eventually realized the potential of undersea warfare were equally varied, as demonstrated by the fact that some of the early exponents of submarine warfare, such as Britain, were initially reluctant to participate in the building of a submarine force. It is this story, regardless of the political connotations, that builds into the dramatic narrative presented in the following pages.

There are two sections in this book, the first of which records some of the most important developments, plus major conflicts and incidents during this most vital period, which saw the emergence of the great submarine fleets in which Britain, America and Germany undoubtedly led the field. Many other nations, such as Russia, France and Italy, followed in close pursuit. This is followed by technical examinations in a directory section, covering boats built or on the drawing boards as works in progress up to 1950, when the first submarines encompassing Jules Verne's vision were already appearing.

# History of Submarines

Submariners will say that, notwithstanding the risks, there is a certain romanticism connected with submarines. After years of trial, error and often fatal attraction, a handful of men emerged in the late 1800s who were clearly making progress towards the realization of the dream, often in the face of adversity and controversy. Traditionalists among the naval hierarchies of several nations – including Britain – fought a bitter struggle to have the submarine banned, but gradually the experimental vessels produced by men such as J.P. Holland and Simon Lake began to show remarkable promise, in spite of the volatile nature and foul conditions inside the boats.

Winston Churchill described submarine development as the most dangerous of all occupations: "Of all the branches of men in the forces there is none which shows more devotion and faces grimmer perils than the submariners." Yet emerging submarine services of nations around the world never lacked volunteers. Indeed, as will be seen, this book tells a very human story, highlighting some of the outstanding events from the very early days to the present, with many tales of courage, drama and quite astounding invention.

LEFT: **In the line of fire: the assassination of Archduke Franz Ferdinand of Austria sent that nation's U-boats, along with those of Germany, into the first major conflict involving submarines, depicted here against a French merchant ship.**

LEFT: **The Confederates' *Hunley*, which sank four times during trials, twice killing all her crew. She was raised each time and sent back into commission, and on her fifth voyage torpedoed and sank the Union warship *Housatonic* off South Carolina, but in doing so went to the bottom with her victim.** BELOW: **David Bushnell's *Turtle* which moved perfectly and with amazing accuracy but failed in its attempts to attach a mine to British ships in New York harbour in 1776. He had more success with his "drifting" mines which harassed the British fleet.**

# Trial and error: the pioneers

Through the ages, inventors and sailors alike sought to find the technology that would allow the construction of boats that could travel under water. Since Alexander the Great was lowered to the bottom of the sea in a glass barrel in 337BC, attempts at such endeavour failed to make any progress until the 16th century. Then, in 1578, the principles of creating a truly submersible craft appeared in the writings of Englishman William Bourne and, although he never actually built one, his ideas were pursued by others.

Among them was Dutch scientist Cornelius Drebbel, who was best known at the time for his perpetual-motion machine, a golden globe mounted on pillars which told the time, date and season. By 1620, his work having attracted the interest of dignitaries across Europe, he began work on designs for his submarine. It was a simple construction akin to a rowing boat with raised sides, made water-tight with a greased leather covering and propelled by four oarsmen. A larger model was later demonstrated to King James I, who was reportedly given a submerged trip along the Thames from Westminster to Greenwich.

The fascination with underwater travel gripped the imagination of the inventors, but naval commanders showed little enthusiasm until the American War of Independence (1775–83), when a young Yale graduate named David Bushnell designed a one-man submarine, which he called *Turtle*. Built from wood in the shape of an egg, his machine was powered by a hand-cranked propeller. Its sole purpose was to damage the British fleet during its blockade of New York harbour in 1776.

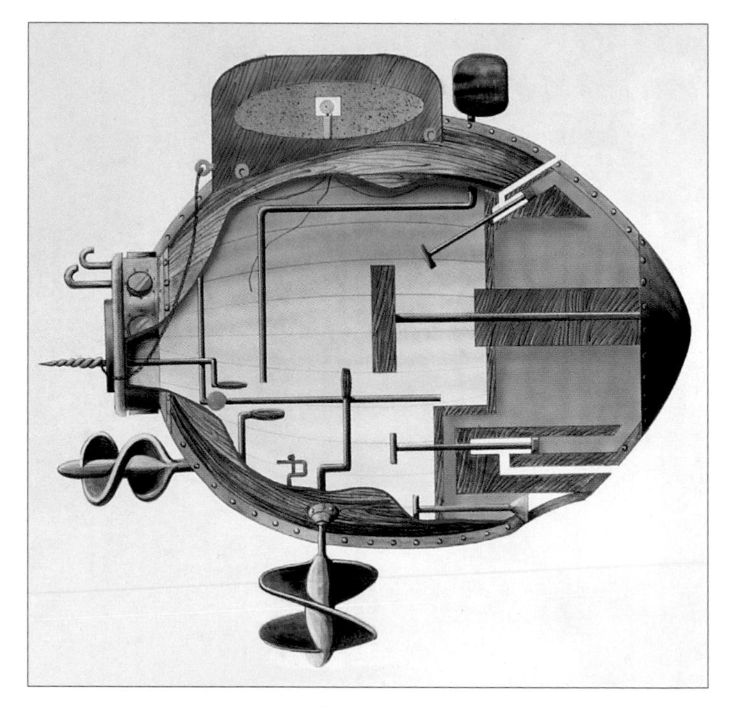

The next step forward came in 1800, when Robert Fulton, a brilliant American artist and inventor, of quite recent Irish descent, built the *Nautilus*, a copper-covered submersible with a collapsible sail. Born in Little Britain township (now Fulton), Pennsylvania, he went to England as a young man to study painting. His thoughts quickly turned from art to draughtsmanship, concentrating on canal navigation and, eventually, submersible boats. In 1796 he travelled to France where, in due course, Napoleon commissioned Fulton to build his submarine, 6.4m/21ft long and shaped like a bullet. Fulton and three mechanics descended to a depth of 7.5m/24ft 7in, and later he added a detachable mine to demonstrate his theory of carrying out clandestine attacks on surface ships. The latter part of the equation failed and Napoleon lost interest. Fulton returned to England and managed to bring his invention to the attention of the British government but the Admiralty rejected his plans with a damning condemnation of submarines that was to remain their policy for the remainder of the 19th century. Back in the

United States, he enjoyed success with steam-powered navigation, and was still working on a new version of his submarine when he died in 1815.

Other attempts at submarine construction met with modest success, including that of a former Germany artillery sergeant, Wilhelm Bauer. He was sponsored to build a submarine, which he called *Fire Diver*, in 1850, but it sank after only two outings. The German Navy was quietly pleased, although Bauer did eventually build an ambitious 16m/52ft 6in submarine for the Imperial Russian Navy, which was known to have completed at least 120 successful dives.

The historical turning point came during the American Civil War (1861–65) when the Confederacy used the first true submarine to sink an enemy ship in war. It was a massive cigar-shaped boat, 18.25m/59ft 11in in length. Named the *Hunley* after one of the three designers who drew up the blueprint for the Confederates' secret weapon, the boat was powered, when submerged, by a manual crankshaft requiring eight members of the crew to propel it at six knots. During trials, the *Hunley* sank four times, twice killing all her crew. Even so, she was raised and sent back into commission. On her fifth journey in 1864, she carried a torpedo to sink the Union warship *Housatonic* off the coast of South Carolina but, in doing so, went to the bottom with her victim. All aboard perished.

The Union, meanwhile, had less success. What is now officially classed as the American Navy's first submarine, USS *Alligator* was designed and built by French diver and inventor Brutus de Villeroi in 1861. Fabricated from riveted iron plates and originally powered by a system of oars, *Alligator* was constructed and tested at Philadelphia on the Delaware River. Later, a hand-cranked screw propeller was added but while being towed to launch an attack on Charleston Harbour in April 1863, the *Alligator* sank somewhere off Cape Hatteras, North Carolina.

TOP: **CSS *David*, a cigar-shaped 15.2m/50ft steam torpedo boat built at Charleston, South Carolina, in 1863 by David C. Ebaugh, was used by the Confederates in a daring attack on Federal ships in 1863.** ABOVE: **A copy of the original blueprint of H.L. Hunley's submarine, which became the first in history to sink a ship.** BELOW LEFT: **The *Pioneer*, a successful hand-cranked submarine built by two wealthy New Orleans engineers in 1862 for the Confederates, but sunk by them to prevent it falling into enemy hands.** BELOW: **Partially submerged incendiary and gun ships attracted the attention of inventors in the 19th century, and this one, designed by Scottish engineer James Nasmyth in 1853, was described as an anti-invasion floating mortar.**

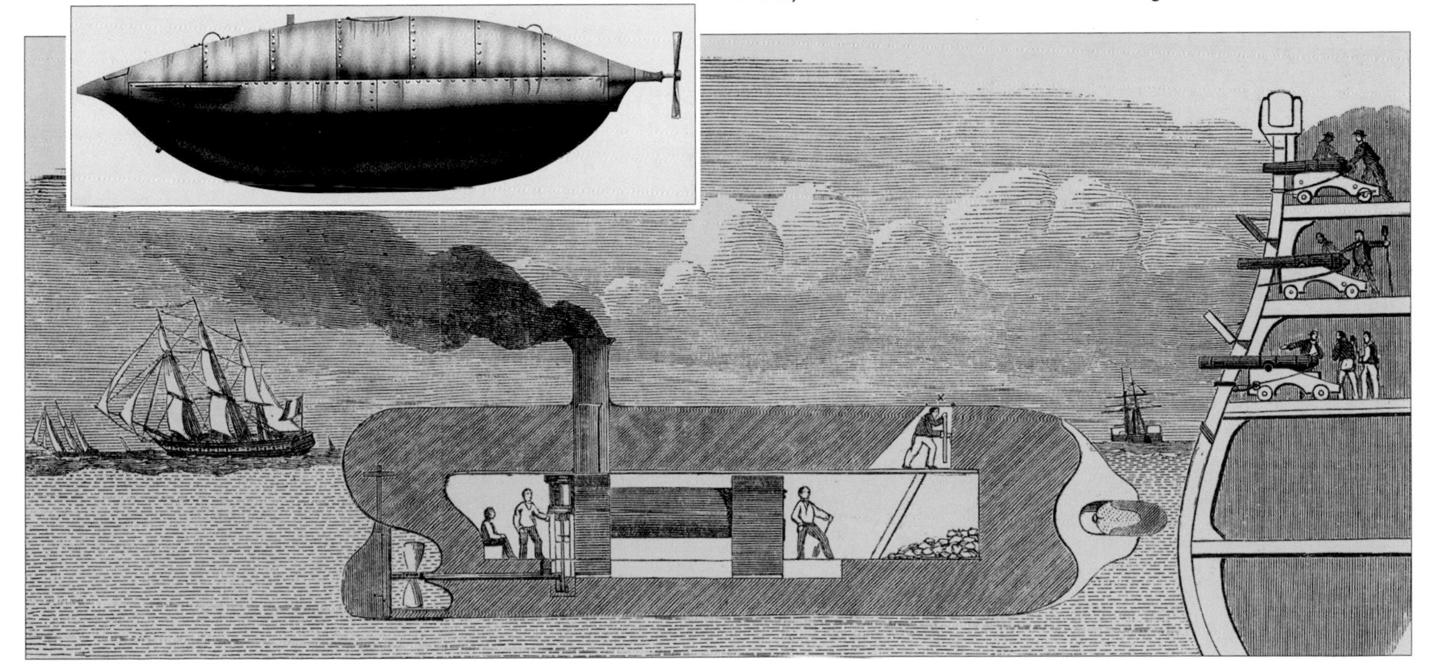

# Closer to the goal

Early setbacks did not deter a worldwide search for a successful design that overcame the most difficult challenge of all, which was to find a workable means of propulsion under water. Another experimental American project involved a boat called the *Intelligent Whale*, a hand-cranked submarine that could stay submerged for ten hours. This was built on the design of Scovel S. Meriam in 1863 by Augustus Price and Cornelius S. Bushnell. The American Submarine Co. was formed to take over production, but this interesting project became bogged down by years of litigation as to who actually owned the rights. When title was established, the *Intelligent Whale* was bought by the US Navy and subsequently abandoned after unreliable trials, eventually being consigned to the Navy Museum, Washington, D.C.

That these difficulties would eventually be overcome was foreshadowed by the appearance in 1870 of the latest novel by Jules Verne, *Twenty Thousand Leagues under the Sea*, the story of Captain Nemo who cruises beneath the oceans in a submarine. It was the inspiration for the Reverend George Garrett, a determined young scholar who had followed his father into the Church but had his heart set on other things. After an education that took him to Trinity College, Cambridge, his interest in mechanical science consumed him. At the age of 26, he produced detailed plans for a bullet-shaped submarine he called *Resurgam* ("rise again"). He raised more than £1,500 and had the submarine built on the banks of the Mersey.

Garrett decided to sail the submarine around the coast to Portsmouth for the ceremonial Spithead Review and set off on December, 10, 1879, with *Resurgam* under tow, crewed by two men while Garrett directed operations from the yacht. The submarine was powered by a coal-fired, single-cylinder steam engine for surface travel, which had to be shut down shortly before the submarine submerged. In theory, residual heat from the engine would keep the vessel under way for about an hour before the need to surface again. It was a dangerous and overpowering contraption, giving off strong carbon monoxide fumes. The crew avoided death from poisoning by wearing a breathing device Garrett himself had invented, although this was not entirely efficient, and lighting was by candles in a craft virtually devoid of instruments.

On February 25, 1880, after a break in the journey at Rhyl, *Resurgam* set off again under tow behind the steam yacht. The weather took a turn for the worse with a heavy swell, and the hawser connecting the yacht to the submarine suddenly went slack. *Resurgam* had gone to the bottom and efforts to find her were unsuccessful. The submarine was to remain undiscovered for more than a century. As for Garrett, he never did get to give the British Navy a demonstration, although a second submarine designed by him and of sleek, revolutionary design did achieve attention.

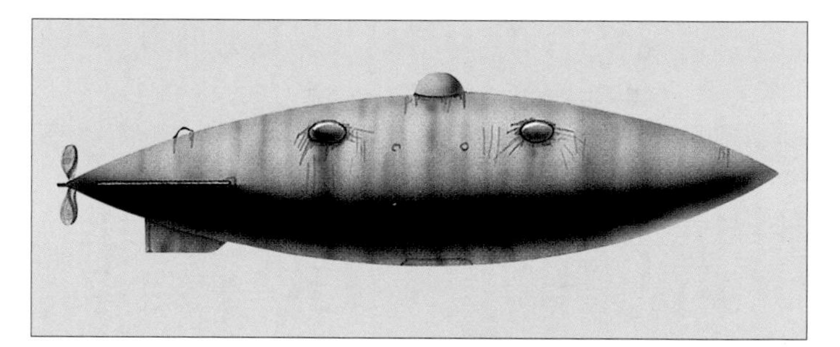

**ABOVE: The *Intelligent Whale*, a hand-cranked submarine built by Augustus Price and Cornelius S. Bushnell in 1862, was eventually sold to the US Navy, subject to the completion of successful trials. However, the first trial failed to impress naval engineers and the project was abandoned without further payment. LEFT: *Resurgam*, the coal-fired and steam-powered submarine built by the Reverend George Garrett in 1879, was originally designed to penetrate the chain netting placed around ships' hulls to defend against attack by torpedo vessels.**

LEFT: **Simon Lake, one of America's first pioneers of submarine development, produced this early boat, *Protector*, in 1901, but it was rejected by the US Navy and sold to Russia in 1904.**

ABOVE RIGHT: **The spacious interior of Simon Lake's first submarine, *Argonaut*, built in 1895 with innovative design features including the ability to stay submerged for 24 hours.** LEFT: **The distinctive wheels of *Argonaut*, added for bottom crawling, and a diver's air-lock hatch, are clearly visible.** BELOW: **Flying the flag, but *Argonaut* was rejected by the US Navy in favour of the *Holland* and Lake began designing submarines for the Austro-Hungarian Navy, although he subsequently built 24 boats for the US Navy during and after World War I.**

Named the *Nordenfelt*, two were built, one in Stockholm and a second at Barrow-in-Furness. Turkey bought the first and awarded Garrett the title of Pasha Garrett. The second, an incredible 36.75m/120ft 7in long and displacing 230 tonnes/226 tons, sank during the delivery voyage to Russia. Garrett, dismayed at his lack of success, emigrated to America to become a farmer, went bankrupt, and ended up in the US army. In 1995 trawlerman Dennis Hunt was fishing off Rhyl when his nets became snagged. A diver friend, Keith Hurley, came out to free his nets and became the first person to lay eyes on *Resurgam* since February 1880.

Back in America towards the end of the 19th century, Simon Lake, a name now synonymous with submarine development in the US, came to the fore – also carrying a copy of Jules Verne's novel under his arm. Lake was born in Pleasantville, New Jersey in 1866, the son of Christopher J. Lake, whose father was the Honourable Simon Lake, one of the founders of Atlantic City. Simon began work at his father's foundry and machine shop in New Jersey, and the prototype *Argonaut Junior*, constructed in wood, was successfully demonstrated. The success led to the formation of the Lake Submarine Company of New Jersey in 1895, which built the *Argonaut*, the first submarine to operate successfully in the open sea in 1898, and which subsequently drew a congratulatory telegram from Jules Verne.

Even so, Russia – not America – gave Simon Lake his first contract to built submarines. The US Navy ran a competition for best design for a revolutionary new submarine. Simon Lake came second, and was snapped up by the Tsar's Royal Navy, moving temporarily to St Petersburg to begin production. The Americans – followed soon afterwards by the British – chose to award their contract to Lake's closest rival, the Irish-American inventor J.P. Holland, who became the primary figure in the next stage of the development of the submarine.

# Underway in the Hollands

When Irish-American pioneer John Philip Holland (1841–1914) won the American competition for submarine designs, the British Royal Navy at last began to take a real interest, given that France, Germany, Russia and Italy were all entering the field. It was Holland's version that became the model for the first submarines to be built for the fleets of both Britain and America, which was rather ironic. Holland was born in County Clare, Ireland, where he joined a religious order in Cork at the age of 17, but other ideas distracted him from a spiritual calling, and he began to draw sketches of submersible boats. He left the order to join the exodus to America, and in 1873 submitted his drawings to the US Navy, which rejected them. However, the American branch of the Irish patriots, the Fenian Society, heard about Holland's work and gave him $6,000 to build two submarines, which they planned to use against the British Navy.

The Fenian brotherhood was founded in New York by veterans of the 1848 Irish uprising to raise funds for the recruitment and training of exiles to fight the British for Irish independence. Holland, desperate for funds, was not in the least restrained by the possibility that his boats might be turned on British shipping because he, too, harboured strong anti-British feelings.

His first submarine was tested with mixed results in the Passaic River in 1878. She sank on her first outing because of loose-fitting plugs and had to be hauled back from the seabed. Holland himself took the controls on his second trip, and it was a successful run, although still leaky. His second submarine, *Fenian Ram*, was launched in 1881. Powered by one of the earliest internal-combustion engines, the boat was 9.5m/31ft 2in long and displaced 19 tonnes/18.7 tons. She went through a series of successful trials in the lower Hudson River and became the subject of close scrutiny by British spies, aware

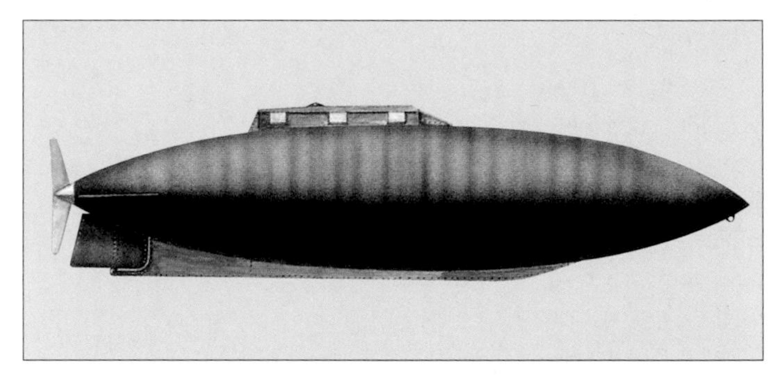

**TOP: The Royal Navy's First Submarine Flotilla, commanded by Captain Reginald Bacon arrived in Portsmouth in 1902, consisting of two completed Holland boats. Five were built at a cost of £35,000 each and Bacon reported: "Even these little boats would be a terror to any ship attempting to remain or pass near a harbour holding them." ABOVE: J.P. Holland's first submarine, the *Fenian Ram*, built in 1878 and funded by a brotherhood of American Irish patriots with the aim of attacking British shipping. The first Holland boats contained many of the original design features.**

of the Fenian connection. Holland's backers, meanwhile, were becoming impatient and demanded action. When he persisted in continuing his rigorous trials they "stole" the boat, which they said was rightfully theirs anyway, and towed it to New Haven, ready to begin blitzing British shipping. After several attempts to master the controls, however, they gave up and washed their hands of submarines.

Holland came to the attention of a businessman named Isaac Rice, a magazine publisher and industrialist who had already established a monopoly in the American storage battery industry. Rice financed the creation of the Holland Torpedo Boat Company and correctly assessed that this time Holland had come up with a winner. The boat was 17m/55ft 9in long and was powered by a petrol engine on the surface and an electric motor when submerged. The United States

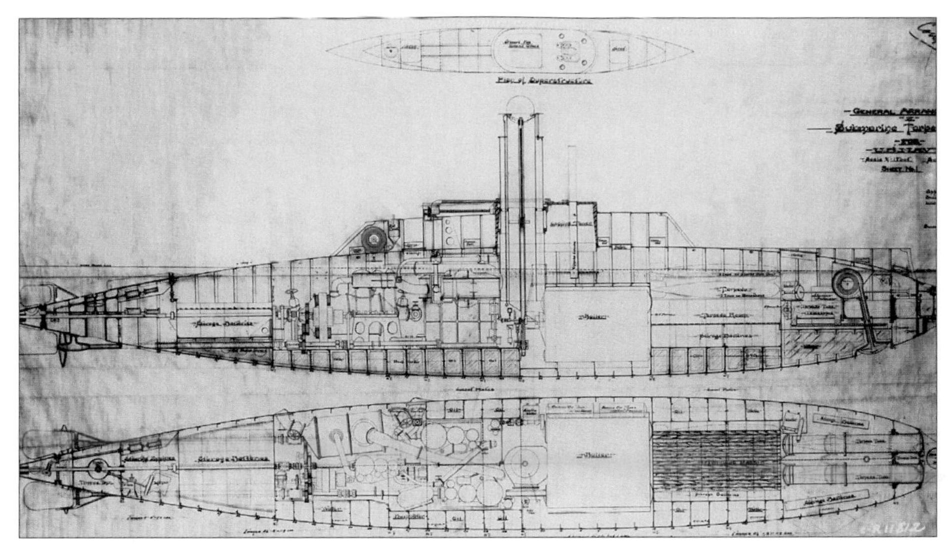

ABOVE: **Drawings for the first of the Holland boats to be bought by the US Navy, the ill-fated *Plunger*, built in 1897 but doomed to failure because US Navy officials insisted on tinkering with Holland's design.**

government finally agreed, in April 1900, to purchase the prototype for $165,000 from the renamed Electric Boat Company, and the craft was commissioned as the USS *Holland*.

Rice then travelled to London, armed with the knowledge of highly successful trials in the USA, where the *Holland* had "sunk" a battleship in fleet exercises, and the Admiralty decided that it, too, must finally join the submarine age. The British agreed to buy five improved Hollands, capable of carrying a crew of seven and able to withstand pressure down to 30m/98ft 5in. Each one cost £35,000 to build, although Holland himself saw little profit from his invention. Rice had taken the precaution of including the designs and patents when he established his company, and eventually Holland resigned. Blocked by legal ties from utilizing his designs again, he died in obscurity while his Electric Boat Company went on to become one of the world's foremost manufacturers of submarines, celebrating its centennial in 1999. Therefore, finally, the British were to get a submarine fleet, although the Admiralty stressed to newspapers at the time that the craft were for experimental purposes only. Lord Selborne, First Lord of the Admiralty, would only go as far as to say in his estimates to the Commons, "What future value these boats may have in naval warfare can only be a matter of conjecture".

This negative view at the Admiralty changed virtually overnight. In July 1901, newspaper reports from France brought a graphic account of dramatic developments. A dummy torpedo struck a French battleship as she left Ajaccio harbour during naval manoeuvres. It had been fired from a brand new French submarine, *Gustav Zede*, which had secretly been sent from Toulon for the express purpose of demonstrating to the navies of the world that the French had exceeded every one of them. For Britain, at a time when France, rather than Germany, was considered an "enemy", this was stern news indeed. The race to form submarine fleets was on, across the world.

ABOVE RIGHT: **The famous image of J.P. Holland in his submarine that provided the prototype for the American and British navies.** BELOW: **USS *Holland*, the 65 tonne/64 ton initiator of the US submarine force, successfully completed trials in 1899.** BOTTOM: **The *Fulton* (Holland VII-type design) built by the Electric Boat Company in 1901. The *Fulton* itself was sold to Russia along with another five, and the US Navy also bought five.**

# Towards world war: the dash to build

To many British and American naval chiefs at the start of the 20th century, submarines were a "cursed invention". The very idea of dark, shadowy vessels lurking unseen beneath the waves, waiting to sink their magnificent and beloved capital ships, was simply unthinkable. Sir Arthur Wilson, Third Sea Lord in the British Admiralty, best summed up their view when he described submarines as being "underwater, underhand and damned unEnglish…certainly no occupation for a gentleman. Submarine crews, if captured, should be hanged as pirates".

The Germans, similarly proud of their naval heritage, felt the same. Admiral Alfred von Tirpitz, said "The submarine is, at present, of no great value in war at sea. We have no money to waste on experimental vessels." The Americans were in no hurry to equip either, in spite of their own history of experiments with underwater warfare. However, the new century was barely months old before the great and the good among the naval fraternity were forced to change their tune. The greater urgency emerged after the French – historically Britain's main enemy before *entente cordiale* — launched a competition for the design of a submarine of 200 tonnes/197 tons, with a range of 161km/100 miles on the surface, and received some startling new ideas, notably Laubeuf's 1899 double-hulled *Narval*, which used steam propulsion for surface work and batteries when submerged.

In 1900, the Americans bought the Holland prototype, commissioned USS *Holland* for immediate operations and began building more. The British introduced their own class, the A boats, the first of which (*A1*) was put through her paces with the five Hollands for the first-ever fleet exercise involving

TOP: **After utilizing the Holland boats as a starting point and thereafter, principally for experimental and training purposes, the Royal Navy initiated its own A class, British designed and built, 12.2m/40ft longer than the Hollands, and at 193 tonnes/190 tons surface displacement was almost a third heavier. Thirteen boats of the class were completed.** ABOVE: **The US equivalent was the Adder class, also an enlarged production version of the original *Holland*, with a more powerful petrol engine.**

submarines, off the south coast of England in 1904. Although based on the *Holland*, *A1* was 12.2m/40ft longer and 203 tonnes/200 tons heavier and was accepted with pride by the Royal Navy. Unfortunately, she did not survive the exhibition. George, Prince of Wales (later King George V), himself a sailor, was aboard for the demonstration, but was called home because of the death of the Duke of Cambridge. Soon after he left, *A1* submerged to continue her trials, and never came back.

The passenger steamer, SS *Berwick Castle*, reported having struck something below the surface. The following day, divers

LEFT: **The French were at the forefront of submarine design. The attractive *Anguilla*, in the Naiade class, was a single hull boat with a surface displacement of 71.6 tonnes/70.5 tons, carried a crew of 12 and had a range of 370km/200 nautical miles.**

ABOVE: **Another French boat that looked well ahead of her time was the *Narval*, built by Laubeuf in 1899. It was the winner of a government-sponsored competition for the best design: a massive 33.8m/111ft long and with a displacement of 170.7 tonnes/168 tons.**

went down at the spot, and found *A1* on her side in 13m/42ft 8in of water. All on board were lost. Five more A class vessels were hit by misfortune, resulting in deaths of crew members. Conditions for the dozen crew members in the early models were dire and cramped. There were no toilet facilities in the early boats. Crews used buckets, which were emptied over the side, and when the air – poisoned by the fumes, toilet buckets and mere body odour – became too overpowering, compressed air bottles were tapped to refresh the atmosphere.

In Britain, the A class was followed by new, improved and larger designs: B, C, D and E classes progressively took the experimentation and improvements at a quite exceptional pace, indeed well in time for the early battles of World War I. Despite this, as late as 1912, senior officials were still rejecting the notion that submarines posed a serious threat to the British fleet.

The era of petrol-driven engines was also thankfully coming to an end. For the British, this was signalled with the arrival of the D-class boats, with *D1* becoming the prototype for diesel-driven power, albeit invented by Rudolph Diesel of Germany. She was also the first British submarine with twin propellers, and her ballast tanks were not inside the hull but fitted on the outside as "saddle tanks". The D boats were also the first submarines to have a wireless, and with their greater endurance level, became ocean-going submarines with a range of more than 4,023km/2,500 miles. A two-pounder gun became standard with *D4*. Eight D boats were in commission by the start of World War I, and along with the other classes, Britain could muster 55 submarines as the conflict began.

France, well ahead in terms of numbers, had a creditable total of 62 very decent submarines. The Germans, latecomers to the race for a submarine stock, had fewer boats than Britain and France. In 1914, they could muster just 28 boats, with a further 17 under construction, but their technology was advancing in new directions that others would copy.

ABOVE: **The race towards submarine power was spurred on by another French development, the *Gustav Zede*, which became the first boat to successfully fire a dummy torpedo, aimed at a French battleship in Ajaccio harbour in 1901.**
BELOW: **The D-letter in Britain's fast track to a major submarine fleet was reached in 1907 and the class had far better endurance than any previous boat. The D class was also regarded as the first UK submarine to have a proper patrol capability.**

# Birth of the U-boat

Germany's early reluctance to join the race in submarine development was only overcome as other nations around the world – France and Britain in particular – began to take up the challenge. In fact, when Germany's history of submarine building began in 1904, it was inspired, curiously enough, not by the country's own navy. Russia ordered three *Karp*-class vessels, which were duly delivered two years before the arrival of the first German submarine, *U1*, which itself was a full six years behind the first British and French boats. The first U (for *Unterseeboot*) boat, also initially designed for coastal work, was larger than British and American A boats, with a double hull and twin screws, and came with a powerful electric motor for faster speeds submerged. The 298kW/400hp engines used heavy oil rather than petrol, although they did have the drawback of emitting very visible smoke and sparks from the upper deck exhaust system. German designers were therefore encouraged to press ahead with diesel-fuelled engines, which they first produced in 1910.

Even so, *U1* produced impressive trial results, and her first major endurance test from the German port of Wilhelmshaven, passing around Denmark and back to Kiel in northern Germany – a distance of 1,087km/587 nautical miles – was noted with concern by competing nations. There was another major difference in the German boats in that from the outset they possessed real fighting qualities, which were quickly enhanced through the early stages of development. It was a capability that Winston Churchill noted on his appointment as First Lord of the Admiralty in 1910: "The whole character of the German

**ABOVE: *U1* was launched at Keil in 1906 to become a trial and evaluation boat for the German Imperial Navy. She was much larger than the British A class, possessed an exemplary electric motor, and had an endurance of 1,087km/ 587 nautical miles.**

fleet shows that it is designed for aggressive and offensive action of the largest possible character in the North Sea and the North Atlantic." Armed initially with a single torpedo tube at the bow, *U2* was further enlarged and came with four tubes, two bow and two stern, ahead of the first British D class carrying four tubes.

*U1*, the pacesetter for fast and capable boats, showed its mettle from the beginning, survived the war, and was eventually moved to a museum in Munich, where she is retained to this day. At the outbreak of war in August 1914, Germany had around 20 operational U-boats assigned to its High Seas Fleet, working from a fortified forward base on the island of Heligoland. When hostilities with Russia began, they combined to form a defensive screen in the North Sea, and when Britain and France declared war on Germany, ten U-boats set out to make a bold surprise move against the British Grand Fleet, at its supposedly secure base at Scapa Flow. This first outing in wartime proved disappointing for the Germans. Two of the boats had to turn back, one was lost en route, and one, *U15*,

**BELOW: The results of what became an all too familiar attack on British merchant shipping in the U-boat campaigns of World War I.**

LEFT: ***U35* was the most successful German submarine in World War I, sinking 224 vessels (excluding warships) during 17 patrols from November 1914 and remaining in action to the very last, when she was surrendered and broken up at Blyth.**

was sunk by HMS *Birmingham*. The remainder failed to score any hits, let alone penetrate the British base, although the latter was finally achieved in November 1914, when *U18* under the command of Captain Hans von Hennig gained entry, only to be spotted and sunk. Although this mass attack on the British fleet achieved little, it came as a wake-up call to all opposing nations, and the British in particular. This was reinforced one month later when *U21*, which had taken part in the move on Scapa Flow, became the first submarine since the American *Hunley* to sink a ship in wartime, sending the British cruiser *Pathfinder* to the bottom with a single torpedo hit.

This was indeed a shocking event, analysed with great concern by the seafaring nations of the world, given that the time which elapsed from the launch of the torpedo to the disappearance of the cruiser was a mere four minutes. Only nine of the 268 crew survived the attack. Undoubtedly the most significant development in submarine warfare to date came on September 22, 1914, when a relatively old boat, *U9*, under the command of Lt Otto Weddigen, sank no fewer than three British cruisers – the *Aboukir*, *Hogue* and *Cressy* – within a 40-minute timeframe, taking down almost 1,400 men. As naval analysts pored over the details, it did not escape anyone's attention that on July 16, 1914, the crew of *U9* had performed the task of reloading torpedoes while submerged, the first time ever that this had been recorded. One further milestone in German submarine history came one month later, when *U17* became the first submarine in history to sink a merchantman. In fact, the boat went on to sink 11 other non-combatant ships.

ABOVE: **The crew of *U9*, then captained by Otto Weddigen, made history by performing the task of reloading torpedoes while submerged. Using this process, Weddigen sank three British cruisers in under an hour in September 1914, with the loss of almost 1,400 men. The captain would himself perish with his *U29* later in the war.** RIGHT: **An interior view of *U9*.**

# The head to head of underwater power

The submarines of World War I were engaged in vast arenas of sea warfare, and the crews of these early, technically deprived small boats were relied upon well into the conflict, taking on challenges way beyond the true level of their capabilities and endurance. Furthermore, the sheer numbers of boats in any fleet did not compensate for deficiencies in the infant submarine technology, although the rate of improvement on that score was dramatic. The British Navy was far greater in both experience and numbers, while the newly equipped German fleet was lean, hungry and thoroughly modern.

ABOVE: **In 1915, a new class of efficient German minelayers came into service around coastal waters. Ironically, the class leader *UC1* was sunk by a mine off Nieuport two years later with all 17 hands lost.** BELOW LEFT: **Some of the targets of extensive German minelaying were the new British E-class boats, which began a daring and highly successful campaign in the Baltic and elsewhere as they emerged to fight in the front line of the war at sea.**

On paper, the submarine stock of Britain and France looked healthy as Europe inched towards World War I. However, well over half the boats were of such vintage as to be virtually useless other than for coastal patrols, and to use them – as the British did – on more demanding tasks presented their crews with exceptional challenges that they met with considerable aplomb.

But other developments were already in the pipeline, and by the end of the war the Royal Navy's submarine fleet would extend across a confusing range of 20 different designs and classes. The star of the British submarine fleet was the versatile E class, with boats being launched by the week to match Germany's massively increased production schedule. Additional types F, G and H classes were scheduled to begin rolling out of the production pens from 1915.

In early action, the Heligoland Bight region remained the key to Germany's ability to set forth, so German commanders sealed the route to the Baltic from the North Sea. This area became an early flashpoint as Britain's C, D and E boats made

LEFT: ***E1*** **was the first of 56 E-class boats which came on stream between 1911 and 1917, built in three groups at an average cost of £105,000 each. Sadly, to avoid capture after a successful campaign,** ***E1*** **was scuttled off Helsinki in April 1918, along with** ***E8, E9*** **and** ***E19.*** BELOW LEFT: ***E32*** **was among just 25 E boats that survived World War I, and was sold in 1922.**

ABOVE: **The famous U-boat poster, one of many produced by Hans Rudi Erdt for the German government in World War I, this one to publicize a film,** ***U-boats Out!*** **on submarine warfare.** BELOW: **Lt Commander Martin Nasmith's** ***E11*****, which began its so-called "career of destruction" against enemy shipping in the Baltic in October 1914. Also in this photograph is Admiral Sir Roger Keyes, who later planned the famous raid on the German submarine base at Zeebrugge.**

constant forays into the region. The problems of going to war in craft with, as yet, still unsophisticated communications soon became apparent. Boats went out of wireless range and, in place, intelligence messages were sent back using homing pigeons, which were usually kept in the forward peak of the boats. Usually, important messages were sent in triplicate – using three pigeons.

Britain's first major sortie was led by Lt Commander Max Horton who went on to become an acclaimed hero in British submarine history. His *E9* sank an old German cruiser, the *Hela*, on the approaches to Heligoland and then scored a direct hit on a German destroyer in the North Sea in mid-September 1914. The euphoria of these first submarine successes was to be short-lived, however. A reversal quickly followed, and another "first" in the history of submarine warfare – the first submarine to be sunk by another submarine. The victim was the British *E3*, which was sighted cruising on the surface in the Heligoland Bight by the *U27* captain, Lt Bernhard Wegener. In the weeks and months that followed, the British 8th Flotilla, operating off the east coast, suffered five further losses, although the German fleet was sufficiently harassed to pull back into the comparative safety of the Baltic Sea.

As well as giving trouble to the German fleet in the Baltic, British submarines had the equally important task of interrupting shipments of iron ore from Sweden to Germany. This was eventually achieved, although in terms of attacks on ships, successes were initially few, partly because the British submarines had been placed under inefficient Russian control. The German propaganda machine complained bitterly of the "underhand and criminal methods of the British pirate submarines" in the Baltic.

LEFT: **The sinking of the passenger liner *Lusitania* by *U20* in February 1915, with the loss of 1,195 lives, brought allegations that the boat laid in wait for the ship as part of a policy of unrestricted warfare, although Lt Commander Walther Schweiger denied it.**
BELOW: **The sinking of the great ship undoubtedly inspired what a German newspaper described as "national pride", and a special medal was cast for the submariners involved.**

# The sinking of the *Lusitania*

Battle for control of the high seas intensified by the day, eventually resorting to what many described as "sheer bloody murder" on February 4, 1915. On that day, Germany declared that the international waters around the coasts of Great Britain and Ireland would henceforth become a war zone. Ships sailing under the British flag would be sunk on sight, and the Germans said they could not guarantee the safety of ships from neutral countries because of the difficulty of identifying flags at a distance or in fog.

On May 7, the threat became a reality. *U20*, under Lt Commander Walther Schweiger, was – according to the German version of events – heading home at the end of a week-long patrol off the south coast of Ireland, having sunk two small steamers and one sailing boat. Schweiger was travelling west from Waterford at 13:20 hours when he saw a very large passenger liner steaming into view, sailing eastwards. It was the *Lusitania*, a magnificent vessel of 31,500 tonnes/31,003 tons, carrying 1,257 passengers and 702 crew. As the liner came closer and closer, she was an unmissable target. Schweiger ordered his crew to battle stations, took aim and at 14:09 hours gave the order to fire.

From a range of less than 730m/798yds, there was no chance of avoidance, no chance of a miss and no chance of escape for the hundreds of men, women and children on board. The torpedo struck amidships. Within 20 minutes, the great liner lurched beneath the waves in a great mass of white foam. Schweiger had sunk the pride of the Cunard Line, and in doing so, 1,195 passengers perished. He watched them as hundreds scrambled to save themselves, and made the following note in his war diary:

*"An unusually heavy explosion takes place with a very strong explosion cloud (cloud reaches far beyond front funnel). The explosion of the torpedo must have been followed by a second one (boiler or coal or powder?). The superstructure right above the point of impact and the bridge are torn asunder, fire breaks out, and smoke envelops the high bridge. The ship stops immediately and heels over to starboard very quickly, immersing simultaneously at the bow. It appears as if the ship were going to capsize very shortly. Great confusion ensues on board; the boats are made clear and some of them are lowered to the water with either stem or stern first and founder immediately. On the port side fewer boats are made clear than on the starboard side on account of the ship's list. The ship blows off [steam]; on the bow the name* Lusitania *becomes visible in golden letters. The funnels were painted black, no flag was set astern. Ship was running twenty knots. Since it seems as if the steamer will keep above water only a short time, we*

LEFT: **Although America had not yet entered the war, the sinking of the *Lusitania* finally brought home to Atlantic travellers the potential of the German U-boat campaign against shipping, given that there were many US passengers among the casualties, whose bodies were carried ashore covered in the national flag.** BELOW: **The famous ship depicted on a postcard of the day.**

*dived to a depth of twenty-four meters and ran out to sea. It would have been impossible for me, anyhow, to fire a second torpedo into this crowd of people struggling to save their lives."*

Many claimed that Schwieger added the last sentence after the voyage, after the world became shocked by the sinking. Indeed, the reaction was one of stunned disbelief, but not in Germany. Naval chief Admiral von Tirpitz received hundreds of telegrams congratulating him. An article in an influential German newspaper summed up the "national pride" with the comment that "the news will be received by the German people with unanimous satisfaction since it proves to England and the whole of the world that Germany is quite in earnest in regard to her submarine warfare". Despite the fact that 124 Americans were among the passengers who died, "riotous scenes" were reported in New York's German clubs and restaurants. Elsewhere, of course, the monstrous crime was universally condemned, and few believed that it was a coincidence that Schweiger just happened upon the *Lusitania* but, more likely, had been lying in wait. That view was reinforced when it became known that the German ambassador to the United States had published a warning to those intending to embark on the *Lusitania*, days before she sailed.

Meanwhile, the deadlocked war of attrition in the trenches and Germany's home-front crisis of shortages of food and supplies through the blockade being waged by the British fleet turned the Germans towards a merciless campaign on the high seas in which submarines were to spearhead a blockade of the British Isles. In spite of the international protests in the wake of the *Lusitania* calamity, Germany began to focus U-boat attention on unarmed merchant shipping in addition, of course, to any British fleet ships they might come upon. In spite of the assurances given to America, in August 1915 *U24* torpedoed and sank the small White Star liner *Arabic* and in early 1916 the French steamer *Sussex*, with the loss of American lives.

ABOVE: **A steamer is torpedoed when, convinced they could starve Britain into submission, the Germans began an all-out U-boat war against merchant shipping. By the autumn of 1916, British and Allied vessels sunk by the Germans reached a staggering 300,000 tonnes/295,262 tons a month, double the totals reached in the summer months.** BELOW: **Nor were attacks limited to below the surface, as demonstrated here, when a U-boat crew launches a surface assault using their powerful deck guns.**

The famous naval confrontation, the Battle of Jutland, waged on May 31 and June 1, 1916, between the British Grand Fleet and the German High Seas Fleet, resolved nothing. British losses in both ships and human lives were greater than Germany's. Even so, Germany's capital ships returned to home ports and did not venture to give battle again during the war. The battle on the high seas was handed over to the U-boat commanders, convinced that they could starve the British into submission.

# Forcing the Dardanelles

Apart from the war in the main European waters, there were battles elsewhere that brought praise and tragedy in equal measure for the British and Allied submariners. The most famous of all came in the Gallipoli campaign against Turkey, as the Royal Navy was tasked with forcing the Dardanelles straits. From the outset, this was to become a life or death mission for many submarine crews when the waterway became bitterly contested as soon as Turkey sided with Germany in the autumn of 1914. In return, the Turks received German military advisers, military hardware and two brand-new ships, the *Breslau*, a light cruiser, and the *Goeben,* a battlecruiser. The two ships were stabled at Constantinople (Istanbul), on the northern coast of the Sea of Marmara, from where the Turks unleashed them to bombard Russian Black Sea ports. In December 1914, Britain's War Council decided to attack Turkey to neutralize the attacks on Russia. Among the advance party in this mission was the submarine service.

For centuries, the Dardanelles had provided the Turks with a natural defence to their capital city. Running south out of the Sea of Marmara past the Gallipoli Peninsula into the Aegean and on to the Mediterranean, the straits are 45km/28 miles long and contained many hazards. Although 6.4km/4 miles from shore to shore at its widest points, there was a rat-run 19.3km/ 12 miles upstream, known as the Narrows, reducing to just over 1.6km/1 mile in width, and notorious for unpredictable currents. The rocky banks of the straits bristled with forts, guns, torpedo launchers and mobile howitzers, while below the water there were floating and fixed minefields and anti-submarine nets. In December 1914, well before the infamous invasion of the Gallipoli Peninsula began, the French and the British sent a large flotilla of ships of all kinds to settle in the wide harbour of Moudros on the Greek island of Limnos. Among them were six submarines, including three small and very basic British B-class boats, *B9*, *B10* and *B11*, all hankering to have a shot at forcing the straits. Lt Norman Holbrook in *B11* was the first to try. On the morning of December 13, 1914, having warned his crew of the dangers, they set off and covered 19.3km/ 12 miles before the boat reached minefields at Sari Siglar Bay. There, the Turkish battleship *Messudieh* was spotted at anchor. Holbrook dived his vessel under five rows of mines and torpedoed the battleship. He then managed to turn around and bring the *B11* safely back, although assailed by gunfire and torpedo boats, having been submerged on one occasion for nine hours.

TOP: **The British and French sent many submarines into the costly attempt to take control of the Dardanelles, in conjunction with the ill-fated Gallipoli campaign. These submarines included some of the earliest and most basic boats, such as the *B11*, which undertook heroic efforts to force a path through the deadly minefields.** ABOVE: **The crew of *B11*, with their skipper, Lt Commander Norman Holbrook – early heroes in the battle for control of the Dardanelle straits.**

The operation brought great acclaim in Britain, and Holbrook was awarded the Victoria Cross, thus becoming the first submariner to be so honoured. The dangers were re-emphasized on January 14, 1915, when the French submarine *Saphir* attempted to get through but was lost with all hands just beyond the Narrows. Naval commanders called for help from the larger E boats, which were promptly despatched from England.

The E boats began arriving in March, ahead of the planned invasion of Allied ground forces scheduled for April 25. The first to try to run the gauntlet of the Dardanelles was *E15*, under the command of Lt Commander Theodore Brodie. In dodging the minefields, his boat was caught by the infamous current and ran aground directly below heavy guns that opened fire immediately. The boat was hit twice, killing Brodie himself, and the crew had no alternative but to surrender.

The next boat to make the attempt was Australia's *AE2*, which began the journey through the straits in the early hours of April 25, the same day that ANZAC troops landed on the beaches at Gallipoli. Lt Commander Hew Stoker, the Irish-born captain of *AE2*, set off steadily on the surface under a moonless sky until searchlights suddenly swept his boat. The shellfire that quickly followed forced him to dive, and as he did so, he ran straight into a minefield. Stoker was forced to keep his boat on the bottom for 13 hours, with barely enough air to sustain life. Finally, he was able to move forward, and at 07:30 hours on April 26, the jubilant skipper signalled that *AE2* had entered the Sea of Marmara. In spite of the somewhat fragile state of his boat, and without a gun, Hew Stoker and his crew immediately began to wreak havoc. Virtually overnight, enemy shipping was curtailed substantially.

Lt Commander Edward Courtney Boyle was dispatched in *E14* to join Stoker. He, too, successfully negotiated the Dardanelles, to enter the Marmara on April 29, and rendezvous with Stoker in Atarki Bay to discuss a plan of action. The following day, *AE2* was shelled by a Turkish torpedo boat and went down, although all hands escaped before the boat sank, and became prisoners of war. Boyle, in *E14*, carried on as planned, carrying out a series of remarkable and daring raids, dodging the Turks hunting for him and sinking several ships before returning to base on May 15, to discover that he too had been awarded the Victoria Cross for his "conspicuous bravery".

ABOVE: **A depiction of one of the most acclaimed events of the Gallipoli campaign, when *B11* negotiated a treacherous minefield five rows deep to torpedo and sink the Turkish battleship *Messudieh*, for which Holbrook was subsequently awarded the Victoria Cross.** BELOW: **The French Navy had numerous submarines in the Allied assault on the Dardanelles, including the *Saphir*, which was lost with all hands in January 1915.**

ABOVE: **Another French casualty was the hefty and very visible *Mariotte*, sunk by murderous Turkish shoreline gunfire after being trapped in anti-submarine nets in the Dardanelles on July 27, 1915.** LEFT: **One of the hero boats of World War I was the Austrialian *AE2*, here in the Suez Canal en route to the Dardanelles, where, after a successful opening sortie, she was sunk by Turkish torpedo boats. All hands escaped, only to be taken prisoner.**

# Stealth and style of the Victoria Cross winners

There were yet other VCs to be won in the dramatic sequence of events that set in stone the stealth, style and drama of British submariners operating in the Dardanelles, and in particular the adventures of *E11*, captained by Lt Commander Martin Nasmith. The exploits of this boat, and others of the period, were to become the stuff of legend, winning worldwide acclaim. Nasmith began his journey through the Dardanelles on the night of May 18, 1915, emerging 16 hours later into the Sea of Marmara. There he and his crew began a remarkable run of success. First, they seized a Turkish vessel and strapped her to the landward side of the submarine to serve as a disguise as she passed shore observers, and using this disguise, *E11* torpedoed a Turkish gunboat and sank several smaller craft. On May 23, a Turkish transport ship, the *Nagara*, which was heading for the Dardanelles, came into view.

*E11* surfaced close by and Nasmith appeared in the conning tower and warned that he was going to sink the ship. When all crew and passengers were safely aboard lifeboats, an *E11* boarding party carried explosives to the Turkish ship. As *E11* sailed away, the *Nagara* blew up in a sheet of flame. She had been carrying ammunition. *E11* now headed for Constantinople itself, where another audacious scheme was hatched. Nasmith planned to deliver an assault on the harbour that supposedly could never be attacked from the sea: his first torpedo smashed into the sea wall and the second sank a moored Turkish gunboat.

The city flew into panic, convinced that a group of British enemy submarines were in the Marmara. Nasmith continued on, diving and surfacing around the approaches to Constantinople, replenishing his own supplies of food, fuel and water from craft his men boarded, and in this way kept on the move for an incredible three weeks before arriving back on June 6, 1915, to the news that he too had been awarded the Victoria Cross. This ongoing campaign was handed over to Edward Courtney Boyle in *E14*, which returned to the Marmara to keep up the pressure, and was joined eight days later by Lt Commander Ken Bruce in *E12*. Nasmith also returned to the fray in July, and literally within minutes of entering the Marmara he had sunk a transport. The next day he sank another transport and then while on the surface shot up a column of Turkish troops, passing along a coast road with his brand new 12-pounder gun. His greatest triumph came just before dawn on August 8, when Nasmith found the Turkish battleship *Hayreddin Barbarossa* in his sights. He stalked the ship for 20 minutes before firing a single torpedo. It scored a direct hit, followed soon afterwards by a massive explosion when the magazine went up. The battleship sank within minutes.

Nasmith and Boyle continued their patrols, sinking ships and bombarding troops on the coast road well into the autumn. They were joined in September 1915 by Lt Commander Archie Cochrane, 33-year-old captain of *E7*, making his second tour of the Marmara. But bad luck dogged Cochrane from the outset. First, he discovered that a second anti-submarine net had been positioned in the Narrows and, as he tried to break through, part of the metal entwined around his propeller. Repeated attempts to free the boat failed, and the Turks, spotting the commotion, began dropping depth mines. When the boat had been submerged for almost 13 hours he decided to surrender to ensure his men's safety. He asked for full astern and managed to surface promptly in the midst of an array of enemy craft waiting to shell him. Amid a hail of gunfire, Lt John Scaife scrambled on deck to surrender the crew. Two motorboats manned by German submarine crews came alongside and took the men off. Cochrane himself

**ABOVE: Light cruiser *Breslau*, one of two warships supplied to the Turks by Germany for the Dardanelles campaign, ran straight into a minefield in the Aegean in January 1918, and was sunk.**
**RIGHT: The wreck of hero boat *E15*, which ran aground en route to the sea of Mamara directly under the fire of defensive batteries, which killed Lt Commander T.S. Brodie and six crew. The remainder were captured and five more died in prison camps.**

ABOVE: ***E11* 1st/Lt Guy D'Oyley-Hughes swam ashore pushing a platform packed with explosives to blow up a railway line that was out of reach of shells.** RIGHT: **Returning safely home, VC winner Lt Commander Martin Nasmith, aged 32, and his crew of *E11* following their action-packed campaign in the Dardanelles, sinking a record 122 enemy vessels, mostly in the harbour of Constantinople, a tally achieved by no other submarine.**

was the last to leave, and as the rescue boats cleared the stricken submarine a time-fuse explosive set by Cochrane before he left was detonated, returning *E7* quickly to the bottom. Cochrane joined Hew Stoker in the prisoner-of-war camp at Yozgat.

Back in the Marmara, Nasmith continued his operations which he now extended to include a number of onshore sabotage missions. He remained operational for an incredible 47 days, a record equalled by no other commander in World War I. During that time, the crew of *E11* sank 11 steamers, 5 large sailing vessels and 30 small sailing vessels. By then, the need for such daring assaults was diminishing rapidly as the land battles around Gallipoli were firmly entrenched in a hopeless stalemate from which there seemed no respite from the daily sacrifice of human life. That one battle zone had engaged more than a million men from the two sides, half of whom became casualties. Early in December 1915, the Allies finally decided to call a halt and to withdraw. In the final days of the campaign, *E14*, now under the command of Lt Commander Geoffrey White, was severely damaged. White gave the order to surface in the hope of making a final dash to safety but immediately came under heavy fire from the shore batteries. With no hope of escape, White altered course towards the shore to give his crew a chance of safety, but was himself killed shortly before *E14* sank. Geoffrey White, who had two sons and a daughter he had never seen, was posthumously awarded the Victoria Cross, received by his widow at Buckingham Palace on July 2, her late husband's birthday. Nonetheless, the statistics for the Dardanelles campaign told their own story: 13 Allied submarines completed 27 successful passages through the straits. Seven were sunk, but the scorecard was somewhat uneven. The Turks lost 2 battleships, 1 destroyer, 5 gunboats, 11 transports, 44 steamers and 148 sailing boats.

ABOVE: **Through the guile and daring of their skippers, the British E class became one of the most successful boats in early British submarine history, with a significant tally that included seven U-boats. They were fast, versatile and safe and much loved by their crews, but the cost was high, with almost half of the 56 E boats built being lost to enemy action.** RIGHT: **Hovering, to keep a low profile.**

# A U-boat in Baltimore

Of all the major seafaring nations of the Western world, the United States Navy was slow in equipping itself with a submarine force, for which it would later become renowned. America did not enter World War I until April 6, 1917, at which time her navy possessed a mere 24 diesel-powered submarines that had been used to patrol the waters off the east coast. Until that date, U-boats had been visiting the US since mid-1916 for trading purposes, as well as operating unhindered in the western Atlantic. In July that year, *Deutschland,* a large German cargo-carrying submarine, broke through the British Atlantic blockade by submerging to avoid British patrols. She arrived in the American port of Baltimore with a shipment of chemicals and dyestuffs, which were traded for a quantity of strategic war materials to be carried back to Germany. The boat, capable of carrying 711 tonnes/700 tons of cargo, made another round trip in November, although the next time the boat travelled the Atlantic, she had been converted to a combatant role.

When America entered the war, US submarines were deployed overseas to the British territory of the Azores and also around the coast of Ireland. Their primary mission was to assist in countering the U-boat threat to Allied shipping, although there were to be no confirmed sinkings of U-boats by American submarines. Back in the territorial waters of the United States, the US Navy began dispatching its submarines as far afield as the Panama Canal Zone and the Philippines, as well as mounting numerous defensive patrols along the east coast, where there was significant concern about U-boat attacks on American shipping.

**TOP: A depiction of one of the epic escapes of World War I, when the passenger line USS *President Lincoln,* which transported 23,000 American troops across the Atlantic, was torpedoed by the German submarine *U90* on May 29, 1918, and sank with the loss of only 26 of the more than 700 personnel on board. ABOVE: An interior view of *Deutschland* at the end of the war, when she was in French hands.**

The threat was very real. Germany had by then created the ultimate World War I U-boat – a true long-range submarine cruiser. These boats were 70.1m/230ft long, with a speed of 15.3 knots on the surface, and an incredible range of 20,326km/12,630 miles at eight knots. They were heavily armed, with twin 15cm/5.9in deck guns for which 1,600 rounds of ammunition were carried along with 19 torpedoes, manned by a crew of 56 with room for 20 more if needed for special missions. Although 47 boats were ordered, only nine were in service before the war ended, six of them deploying to the east coast of the United States, where they laid mines and sank 174 ships, mostly smaller vessels without radios which could

ABOVE: **The compartmentalized *Deutschland*, one of the first of the UA-class boats. These appeared in 1916 and were considered to be the ultimate long-range submarine cruisers. They were 70m/230ft long, 1,524 tonnes/1500 tons, had a speed of 15.3 knots on the surface, a range of 20,326km/12,630 miles at 8 knots, and carried a crew of 56. *Deutschland*, specifically built as a blockade-breaking civilian cargo submarine, with a carrying capacity of 711 tonnes/700 tons, caused an uproar among the Allied nations in Europe when the US government allowed her access to US ports, carrying dyestuff and gemstones to America and returning with a cargo of nickel, tin and rubber. She made two trips in civilian mode before being converted for hostilities.** BELOW: **Fully flagged and impressively majestic, the largest of the German submarines of that era, pictured in civilian mode, was clearly a model for future developments.**

neither be warned or give warning. They proved emphatically that submarines could operate with ease 6,437km/4,000 miles from their home base.

By comparison, the dozen submarines that America sent to the Azores and Ireland had little or no effect on the outcome of the war, in that they existed mainly to provide an anti-submarine patrol for the trans-Atlantic convoys, using L-class boats that were no match for German submarines. One statistic alone proved the point: the L boats took almost two-and-a-half-minutes to dive, whereas the U-boats were out of sight in less than half a minute.

One of the most successful boats in the cruiser flotilla was *U151*, which left Kiel in mid-April 1918 and mined the entrances to Chesapeake and Delaware Bays and severed several telegraph cables near New York. During a career of four major patrols, the boat sank 51 ships totalling 140,503 tonnes/138,284 tons – excluding warships. Of these, 23 totalling 61,979 tonnes/61,000 tons were sunk off New Jersey and North Carolina in a seven week onslaught in the summer of 1918. The most notorious became known locally as Black Sunday – June 2, 1918. On that day, *U151* destroyed six American vessels, including the passenger liner SS *Carolina* en route to New York from San Juan with 217 passengers and 113 crew aboard. Although the U-boat captain observed the rules, and allowed all aboard to take to the lifeboats before sinking her, 13 passengers were drowned.

These seemingly fearless German long range boats also sank or damaged several major American warships. It was a stark lesson for the Americans, as the Germans had convincingly demonstrated that modern submarines could operate with ease over transoceanic distances.

When the war ended, the US Navy had 74 submarines in commission, with another 59 under construction. Except for two submarines sunk in accidents, the US Navy lost no submarines to enemy action, and by early February 1919, all the boats that had served in the Azores and Ireland had returned to the United States. America knew by then that her submarine resources were insufficient compared to those of Europe, and set about developing a world-class force that began to emerge in the late 1920s and beyond.

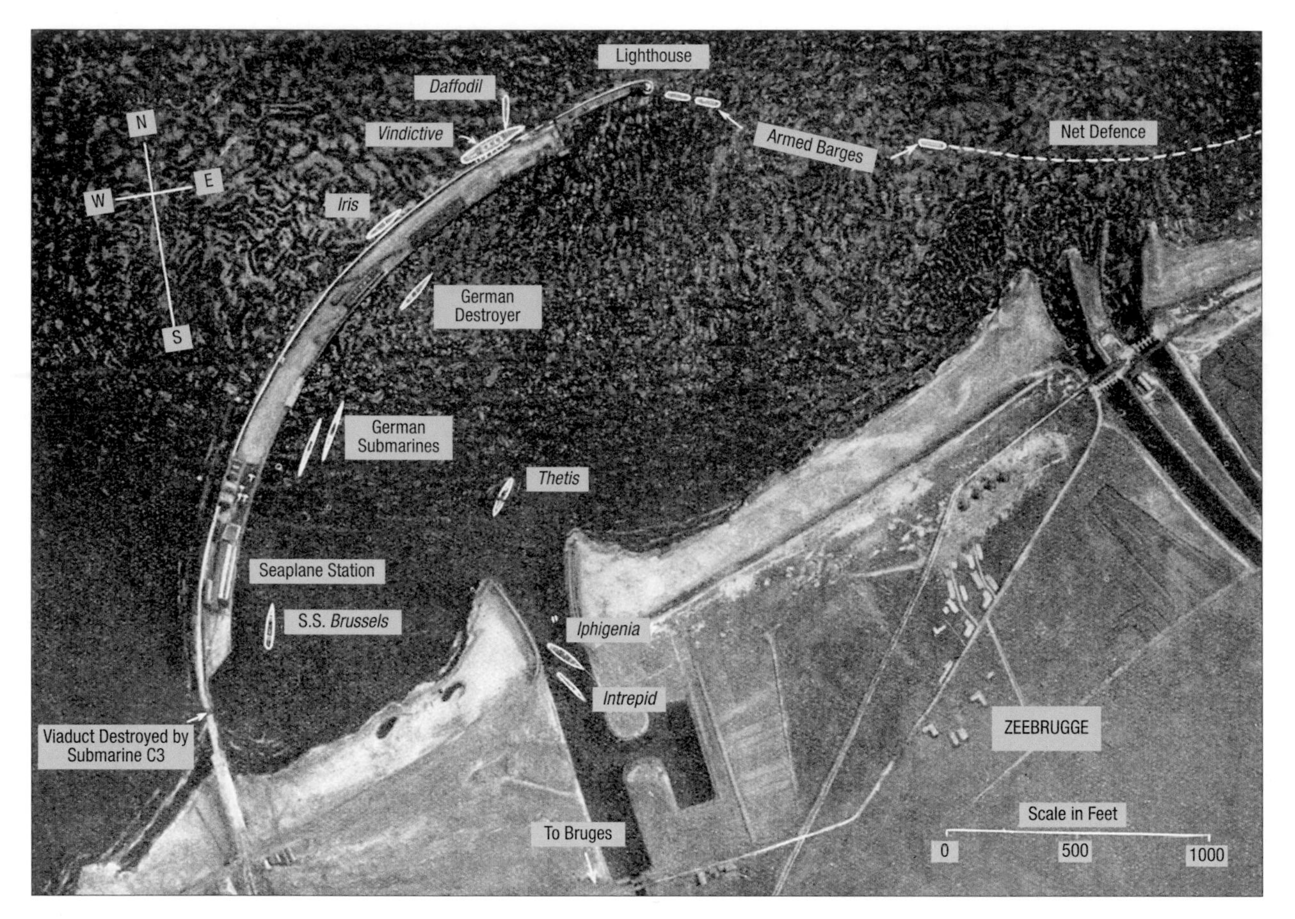


LEFT: **An aerial view of the waterway to be attacked and blocked.**

# The last vital, courageous act

Many of the U-boats that caused so much damage in the Atlantic had been based at Bruges, in Belgium, following its capture by German land forces. It was because of this forward position from which the U-boats were able to operate that the system of convoying fleets of merchant ships was introduced. Warships stood guard, seaplanes spotted the submarines and depth bombs or charges were available for destroying them. By the beginning of 1918, the convoy system allowed the Allies to begin to regain the momentum, but the U-boat base remained a major thorn in their side. Finally, in the spring of that year, one major, daring operation was planned to directly attack the U-boat campaign at its source.

The U-boats' safe haven was strongly defended, allowing boats to come and go with comparative ease. Their hideaway was 13km/8 miles from the open sea and accessed through two canals, one via Ostend and the other through Zeebrugge. A joint operation was planned utilizing air, land and sea units to block the canals and force the U-boats to use the more dangerous northerly route, where the British could attack them. It was proposed that old ships could be sailed to the canal entrances and then sunk, a task that the Royal Navy undertook with relish with the Royal Marines in a large military operation involving 1,760 men and numerous vessels.

A crucial element of the main operation was handed to the 6th Submarine Flotilla. It entailed blowing up a viaduct connecting the Zeebrugge mole to the mainland to stop troop reinforcements being brought to the scene at the time of the raid. Two older-type submarines, *C1* and *C3*, were nominated for duty. Six tonnes/5.9 tons of explosives were packed into the bows, and it was planned to ram the submarines under the viaduct using an automatic pilot. In theory, the crew would make a quick getaway in a motor skiff 1.6km/1 mile from the target, and they would hopefully be well clear when the time-fuse blew everything up.

ABOVE: **The vintage submarine *C3*, one of the two boats packed with explosives to be rammed into the viaduct while the block ships were being put into position. Unfortunately, the second boat, *C1*, broke down and it was left to some heroic work by the five-man crew of *C3* to ensure her deadly cargo was positioned and time-fuses set before they made their escape.**

Only unmarried volunteers were accepted, and the crew was stripped to a bare minimum of two officers and four ratings. Lt Aubrey Newbold was already captain of *C1*, while Lt Richard Sandford, who had just reached his 26th birthday, took

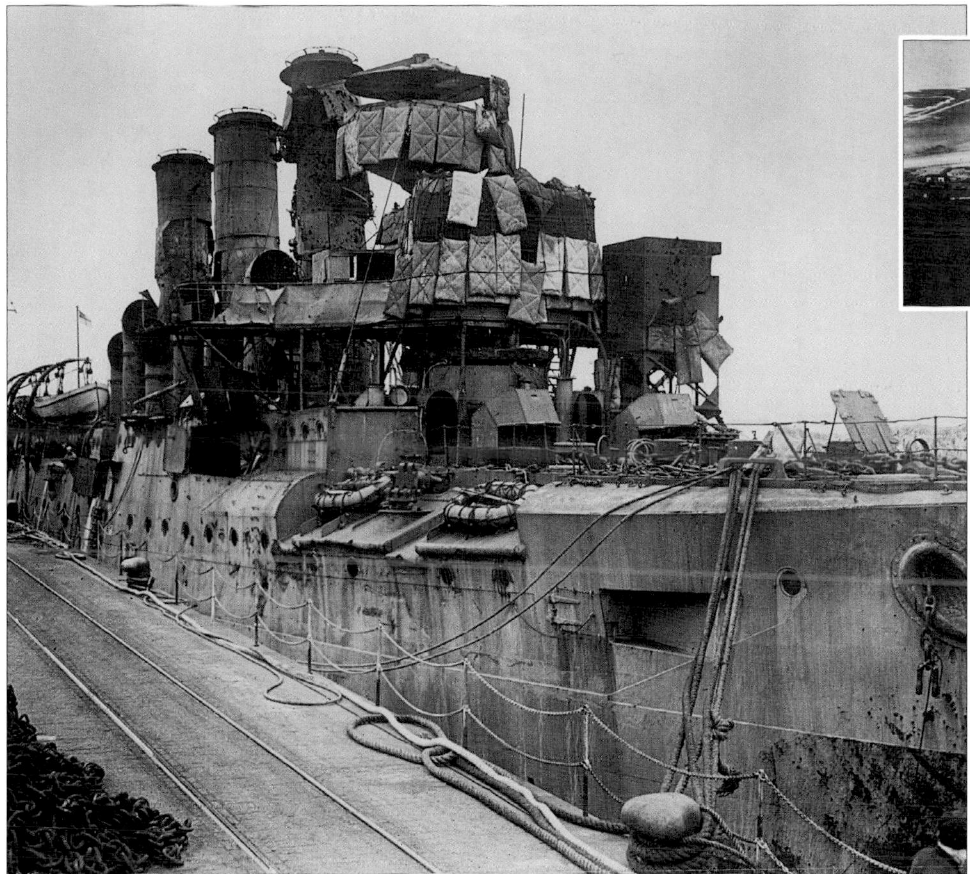

LEFT: **The ancient cruiser HMS *Vindictive* was one of the principal ships carrying the diversionary raiding parties of Royal Marines who swarmed ashore at Zeebrugge to engage the German defenders until the block ships had been manoeuvred into position and scuttled (ABOVE). These were to block the entrance to the harbour used by U-boats for their forays into the Atlantic. While the Marines dashed ashore, the ship's gunners bombarded the German emplacements on the mole. Even so, she remained an easy target for the incoming shells and took a number of severe hits. *Vindictive* managed to get away when the task was complete, albeit with heavy casualties.**

command of *C3*. The task force moved out after dark on April 22, with the two submarines towed by destroyers towards a rendezvous point 8km/5 miles from the viaduct. *C1* never made it, having been delayed by mechanical trouble en route, and Sandford took *C3* into attack alone, but made a significant deviation from the original plan. He was to have set the submarine on automatic pilot for the final stages of the journey, thus allowing him and his crew to abandon the submarine at a safe distance. However, having lost his companion boat, he decided to manually steer right to the last. German gun emplacements were firing all around, while star shells bursting directly overhead brought fake daylight for the gunners to take a range and fire. Sandford held course and made it to the target, ramming the bow of the submarine into the side of the masonry. Overhead, there were howls of laughter as German troops mistakenly believed a British submarine had become stuck.

Even so, as Sandford led his five crew members out after setting the time-fuse, the Germans opened fire. Their little motor skiff was badly shot up and the engine was useless, so they had to row. As they pushed off, two of them were hit, then Sandford himself took two bullets. Four of the six were wounded, but they managed to keep rowing until they were rescued by a picket boat, driven by Sandford's brother, from their destroyer. As they climbed aboard, the sound of a terrific explosion from the place they had just vacated confirmed that the job was done – and, indeed the viaduct was missing a very large chunk in the middle of the run. All who took part in the operation received medals, and Sandford was among eight officers involved who were awarded Victoria Crosses. His citation read:

ABOVE: **The results of the explosion can be clearly seen – a gaping stretch of broken masonry after the submarine exploded.**

*"He eagerly undertook this hazardous enterprise, although well aware (as were all his crew) that if the means of rescue failed and he or any of his crew were in the water at the moment of the explosion, they could be killed outright by the force of such explosion. Yet Lieutenant Sandford disdained to use the gyro steering, which would have enabled him and his crew to abandon the submarine at a safe distance, and preferred to make sure, as far as humanly possible, of the accomplishment of his duty."*

This was represented at the time as a tremendous British victory by Allied propaganda with the consequence that its devisor, the Navy's youngest admiral Sir Roger Keyes, was ennobled. Conversely, the Germans cited it as a demonstration of their success by holding the two ports. In reality, the Zeebrugge raid did not hinder German operations from either port for more than a few days. Some 500 British casualties were incurred during the operation, of which approximately 200 were fatalities.

TOP: **The image that came to signify the horror of the mounting U-boat attacks among the Allied fleets, a fate that befell 6,596 merchant ships totalling 13,006,146 tonnes/12,800,733 tons during the war years.** ABOVE: **To the Germans, their submariners were heroes, and there are numerous memorials to them, such as this one, which mentions the U-boat flotillas in the Aegean and Mediterranean seas.**

# The final reckoning

The threat of what could have been a massive U-boat offensive against the Allies was only revealed at the end of the war. While talk of mutiny and rebellion among the German Navy matched the mood of many of the nation's citizens, the submarine service remained loyal to the last, with 176 U-boats still in operation and another 200 or so still under construction. However, there was no role for them, at least not in the foreseeable future. Under the Treaty of Versailles, Germany was hammered into straitjacket surrender terms, and in addition was to have her fleet confiscated and her government was barred from keeping, buying or building any submarines whatsoever.

The Germans were forced to hand over the bulk of their ships to the Allies, comprising 10 battleships, 17 cruisers, 50 torpedo boats and 176 submarines. Most of the fleet, with the exception of the submarines, was interned at Scapa Flow in November 1918, where many were scuttled by their own crews under the very noses of their British guards. All operational U-boats were delivered into the hands of the Royal Navy initially at the port of Harwich, temporarily giving the British the largest submarine fleet in the world. In fact, the spoils were shared among the Allies and some even ended up in Japan, allowing naval architects and scientists to copy the technology.

Most were eventually scrapped, and 250 damaged or unfinished U-boats were also destroyed where they lay in the German shipyards. It seemed an appalling waste, but in reality this was trifling when set against the cost of the war itself to all the belligerents, amounting in total to about $186 billion. Casualties in the land forces amounted to more than 37 million men, with almost 10 million deaths among the civilian populations caused indirectly by the war. Famously talked of as "the war to end all wars", which would bring permanent world peace, quite the reverse would result.

From the British standpoint, the only salvation was that the nation had not fallen to defeat at sea at the hands of the U-boat commanders, although it had been a very close contest. Even so, the nation's near bankruptcy as the war ended more or less forced the British government to embark on a major culling programme of its own, scrapping and selling dozens of submarines, many of which had seen little service and some not even commissioned. At the end of the war, Britain's submarine stock included 10 out of the original 13 A boats, 9 of 11 Bs, 27 of 38 Cs and 3 of the 9 Ds remaining in service. All were scrapped, as were the following:

**E class**: These boats had seen terrific service during the war years but were all destined for the breaker's yard. Twenty-seven

had been sunk in enemy action, mined, scuttled or lost for reasons unknown. The remaining 30 were all scrapped by 1924.

**F class**: Only three were built; all were used for training and then scrapped.

**G class:** Designed initially as an overseas patrol boat, with powerful engines and long endurance submerged, the G class also had the distinction of being the first to carry a 533mm/21in torpedo as well as two 457mm/18in tubes on both bow and beam. Fourteen were built, and four were lost during the war, including *G67*, the last submarine lost to enemy action, sunk in the North Sea on November 1, 1918. The remaining ten were to be sold or scrapped.

**H class**: Forty-three were built between 1914 and 1918; this was another workhorse of the period. Eight were lost, 18 were scrapped or sold in the 1920s and the remainder were kept in service for several years hence; some still active in World War II.

**J class:** Seven were launched, one lost and the remainder transferred to Australia in 1919.

**L class**: Another major production schedule produced this class from midway through the war. The design was based on an elongated E class; 34 were ordered, although only 27 were commissioned. They were of variable designs to provide boats for basic torpedo armament, as minelayers, and later all carried a 102mm/4in gun mounted on the superstructure; three were lost (two accidentally) during the war and another sank with all 40 hands in 1924. The rest remained in service until the 1930s, when most were sold, although a handful was still in service at the beginning of World War II.

**R class:** Ten were launched in 1918, but few were commissioned before the war's end. They were notable for their very attractive streamlined hull design, which produced a fast underwater speed capability. They were designed specifically as attack boats and had a huge firepower of six 457mm/18in torpedo tubes. Since most of them were launched when the need for attack boats had virtually passed, they saw little in the way of active service and were all scrapped in 1923. Two more, known as the second R class, with various amendments, were built in 1930.

ABOVE LEFT: **Under the terms of the Armistice at the end of World War I, Germany was required to surrender her submarines to the Allied nations. Britain received a large number at the east coast port of Harwich and these two, *U124* and *U164*, were almost brand new.** ABOVE: ***U117* was among the boats surrendered to the USA and was subsequently used for naval exhibitions along the Atlantic coast before she sank in June 1921 off Virginia during tests.**

ABOVE: ***U120* had been operational for only two months prior to the end of the war and was one of the boats surrendered to Italy for inspection by their own designers and engineers before being broken up at La Spezia in April 1919.** BELOW: **The continuing stream of forlorn U-boat commanders and skeleton crews who surrendered vessels allotted to Britain at the end the war, coming ashore at Harwich in 1919.**

# The "jinx" on the K boats

Among the boats retained for post-war service by the Royal Navy were those in the experimental K class, a steam-driven fleet submarine, hugely impressive in appearance and, at the time of their first commissioning in 1916, the largest and fastest submarines in the world, bar none. Unfortunately, their performance and stability did not live up to the great expectations surrounding them, and after a succession of disasters, the tag of "jinxed" was soon assigned.

Their design called for steam engines, which had never been completely successful in submarines, not least because they required funnels. They were 103m/337ft 11in in length, with a surface displacement of 1,850 tonnes/1,821 tons and 2,450 tonnes/2,411 tons submerged, driven by a pair of massive steam turbines powered by oil-fired boilers which could produce a speed of 23 knots on the surface and 10 knots dived from four electric motors. There were eight 457mm/18in torpedo tubes on the bow and beam and a crew of 55.

However, steam power had a number of drawbacks. The boilers had to be shut down on the command of "Dive!" a process that took up to five minutes. The funnels had to be retracted into their housings and covered by watertight shutters. Four mushroom-shaped ventilators that took heat from the boiler room were also retracted and sealed. It was perhaps a prophetic omen that the first of the class to be accepted into service almost changed the course of history in a serious way. The future King George VI, then a youthful naval officer in the Grand Fleet, was having a personal demonstration run in the vessel when the boat went into an uncontrolled dive, and ended up bow down on the muddy bottom in 40m/131ft 3in of water off Portsmouth. It took 20 minutes of delicate manoeuvres before Commander Leir managed to extricate his boat from the mud and return the future king to safety.

*K6* also hit problems during her diving trials. Having submerged during tests at Devonport, the boat simply refused to return to the surface after the ballast tanks had been blown. The 50 men on board were trapped for two hours before the trouble was traced to a fault in the compressed-air system. Next, *K4* ran aground at Walney Island during surface trials, and then *K11* suffered generator failure when she too shipped water through the funnel intakes during trials in heavy seas.

On January 29, 1917, *K13* set off from the Clyde heading for Gare Loch with a crew of 53 complemented on this day by civilian passengers, mostly VIPs from the shipbuilders, 82 people in all. Barely had the boat reached the dive depth when the boiler room began to flood. *K13* dropped to the bottom with a bump. As the captain picked up the voice-pipe to give the order to stop engines, water spurted out, which meant the engine room was totally flooded. Then flames erupted on the

**ABOVE: *K22* (ex-*K13*) lowering her steam-turbine funnels prior to diving, a procedure that took up to five minutes. She sank on her maiden voyage with 82 crew and civilians aboard. Only 47 survived. BELOW: *K6* also sank on her diving trials, and went to the bottom where she stubbornly remained for two hours until a fault was traced.**

ABOVE LEFT: ***K2* suffered an explosion in the engine room just after she submerged and only the quick thinking of skipper Lt Commander Noel Laurence in surfacing immediately saved her from disaster.** ABOVE: **K-class ratings were confronted with endless problems and dangers in these boats. Many of their colleagues did not survive.**

control switchboard, which the men beat out with their bare hands. The boat was immovable, the flooding was becoming more serious by the second and there was no means of contacting the surface. Two men courageously went out through the conning tower to raise the alarm, one of whom died in the attempt. Several hours passed before a rescue operation could be mounted. A head-count of survivors on the boat revealed that 31 on board were already dead.

Further deaths occurred before air pipes were fed into the boat from the surface after 35 hours, but it was another 22 hours before wire ropes had been looped under the bows to raise the boat and holes were cut through the hulls with oxyacetylene equipment to free the remaining 47 still alive on board. A subsequent court of inquiry heard that an inspection had revealed that the ventilator doors were open and that the instruments showed it. The boat was salvaged and was quickly recommissioned as *K22*.

Further dramas were soon to follow, including one of the most horrific accidents in naval history. On February 1, 1918, two submarine flotillas, the 12th and 13th, made up entirely of K-class boats, nine in all, were assigned for exercises with the fleet in the Firth of Forth. As they approached the estuary, the 13th Flotilla had to take avoiding action when a flotilla of destroyers crossed their path. *K14* was rammed by *K22* (the salvaged *K13*). Both submarines were damaged, and as she battled to stay afloat with flooding problems *K22* was hit squarely by the battlecruiser *Inflexible* and had to be taken under tow.

In the chaos that developed, *K17* was rammed and it vanished within seven minutes. Miraculously, all 57 crew managed to escape and were splashing about in the water waiting to be rescued when three destroyers steamed directly towards them unaware of their presence and ran over them. Only eight survived. In the continuing confusion, *K6* rammed *K4*, which sank instantly, and her entire crew of 55 went down with her. The misfortunes of these magnificent boats continued and, in all, 16 were involved in major accidents, and eight finished up on the bottom of the sea before the class was scrapped in 1932. Over 300 men lost their lives in K-class or converted K-class boats – not one of them to enemy action.

BELOW: **The K class was in fact an imposing looking vessel, but designed in haste during World War I, when naval intelligence learned that the Germans had launched a U-boat capable of 20 knots on the surface. The K boats were capable of 23 knots.** RIGHT: **The familiar bulbous nose of *K12*, one of the lucky ones of the class, surviving a near-miss during exercises with the fleet.**

# Big guns and aircraft carriers

Innovation in the design and use of submarines has been constant over the years, and none more so than in the early days when experimentation and high-risk projects were quickly coming off the ideas boards. Two in particular dominated the thoughts of many design enthusiasts in the British Naval hierarchy, especially during the war years in Europe, when competitive military might became the mother of invention.

The first aircraft to be used as bombers, for example, instantly caught the imagination of many. Apart from the obvious notion of surface carriers, the prospect of aircraft launched from submarines became an immediate prospect, as indeed did the use of big guns in the style of surface warships. In August 1915, Admiral Sir John "Jackie" Fisher, the instigator of the Dreadnought class of battleship, voiced his theory of having a submarine equivalent with a big gun that could surface amid flotillas of enemy vessels, perform untold damage and vanish again.

Four boats from the K class, *18*, *19*, *20* and *21* were selected for conversion into M class to be reconfigured with revolutionary features that were to make them more like submersible battleships. They were to be armed with a 60-tonne/59.1 tons, 305mm/12in gun, which had a much greater range than any torpedo but could be fired only from a depth of 6m/19ft 8in, hardly a concealed position, and reloading had to be carried out on the surface. Although approval was given for the conversion in 1916, delays and Navy red tape meant none was ready in time for war service and only three were completed. They became *M1*, *M2* and *M3* and looked exceedingly menacing but never fired a shot in anger. They were rendered obsolete by the 1921 Washington Disarmament Treaty, which limited the size of submarine-mounted guns to 216mm/8.5in.

Nor did the change of prefix rid the boats of the jinx of the K class. While submerged, *M1* sank after a collision with a Swedish coaster, SS *Vidar*, off Start Point on November 12,

**TOP: A famous image of *M2*, the first British boat to be turned into an aircraft carrier, although ambition proved more hopeful than reality in their use, and the jinx of the K class, from which they were converted, continued. She was lost with all hands. ABOVE: The speed of take-off had to be brisk with the engine racing, as demonstrated here, to achieve lift to avoid dumping into the sea, as often happened.**

1925, with the loss of her crew of 68. The wreck was found in 1999 by a sport diver 35 miles south-east of Plymouth.

*M2*, meanwhile, had a hangar fitted with a gantry to lift a light seaplane, the Parnell Peto, with an endurance of two hours, aboard to become Britain's first submarine aircraft carrier. She too suffered the K-fate. On January 26, 1932, the boat was seen to nose-dive at an angle of about 45 degrees by a passing cargo steamer whose captain did not know anything about submarines and did not realize the significance of such a dive. Only when newspapers carried the reports of a missing submarines did he come forward, but by then the 67 crew and two RAF airmen had already perished. Before long, the submariners who knew of the K-class stories were pointing out that *M2* was commissioned exactly 13 years earlier.

Only *M3* saved the reputation of the class, operating as an experimental minelayer stowing 100 mines on rails inside a free-flooding casing. The mines were laid out over her stern by means of a chain conveyor belt.

The death of *M2* finally brought an end to British attempts to launch an effective submarine aircraft carrier, the Royal Navy

ABOVE: **The Americans' *S1* submarine became the experimental platform for their hull-mounted collapsible seaplane, here the XS-2 all-metal version. After surfacing, the aircraft could be rolled out of its pod, quickly assembled, and launched by balancing the submarine until the deck was awash.** LEFT: **The Japanese produced some of the best examples of submarine aircraft carriers, notably the *I400* with a hangar carrying three catapult-launched Aichi M6A seaplanes.** BELOW: **The Japanese submarine-launched plane E14Y, the only aircraft to drop bombs on the United States mainland in World War II.** BOTTOM: **Britain's *M3* which was converted to carry a 305mm/12in gun, and later became a minesweeper.**

constructors having failed in earlier attempts using Sopwith Schneider floatplanes attached to a modified *E22*. This submarine was converted in 1916, with the intention of intercepting German airships over the North Sea. The distinct disadvantage of this first carrier was the boat could not submerge with the aircraft aboard. Only one successful trial was carried out before the submarine sank.

Other navies continued with the carrier experiments. France, Italy and America all produced submarine carriers but none matched the extensive work in this area carried out by the Japanese who began building aircraft carrying submarines in 1925. Most of their early boats were classified as scouting submarines, *B1* Type, of the I15 class, displacing 2,626 tonnes/2,584 tons submerged and 108.5m/356ft in length. Powered by twin diesel engines and electric motors driving two propeller shafts, the *B1* type boats had a cruising range of more than 22,530km/14,000 miles. They were also heavily armed, and thus doubled as attack boats, carrying 17 torpedoes and sporting a 140mm/5.5in deck gun. The crews were comprised of 97 officers and enlisted men.

The floatplane carried by the *B1* model was collapsed into a water-tight hangar installed forward of the conning tower. Two launching rails extended from the hangar to the bow, from which the reassembled aircraft would be catapulted into the air by compressed air. The returning aircraft had to land on the sea and was recovered using a retractable crane. The Uokosuka E14Y aircraft had an operating range of about 322km/200 miles.

The United States Navy, like the Royal Navy, had no such success with their attempts to launch a submarine aircraft carrier. Their experiments began in 1923, using an S-class submarine with an MS-1 seaplane on deck. The aircraft was to be stowed, disassembled, in a cylinder on deck while the

submarine was submerged. It proved to be a totally impractical idea and was abandoned in 1926. Meanwhile, the Japanese proceeded to enhance and advance their own technology in this area, and by World War II they had 47 boats capable of carrying seaplanes. *I14*, for example, was fitted with a hangar capable of housing two aircraft, and the gigantic I400-class boats could carry three. All were at sea preparing for an attack on the US fleet anchorage at Ulithi Atoll, when the Japanese surrendered.

TOP: **The giant, USS *Nautilus*, one of the *Narwhal*-class submarines, the fifth ship of the US Navy to bear that name, although she was designated *V-6*. Commissioned in 1930, she won a presidential citation for aggressive patrolling and 14 battle stars in World War II.**
ABOVE: ***Nautilus* under construction.**

# Giants of the deep

By the mid-1920s, it was clear that Britain had made a rod for her own back with rapid developments of submarine technology matched and copied by other nations, now sharing the benefits of the technology gleaned from captured U-boats. Having come through the devastating experience of having so much of the nation's merchant fleet wiped out, the British naval chiefs went through a reappraisal of their own and many reverted to the original thoughts of the traditionalists – that submarines were a cursed invention. In attending the 1921 post-war Washington conference of the five naval powers – the others being France, the United States, Italy and Japan – the British representatives shocked everyone by proposing an international ban on submarines.

The idea was rejected out of hand by the other nations, understandably since most – especially America and Japan – were already embarking on a massive submarine development programme. What the British and the Germans had produced during the war years, however, amounted to a catalogue of designs and ideas that would provide base plans for the construction of bigger, faster, more potent vessels.

The designs of the two senior combatant nations set the standard for progressive and bold experiments that resulted in the production of the world's largest ever submarine, *X1*. The road to that development emerged directly from the fear of what Germany possessed, and vice versa. The British move towards "bigger and better" began first with the production of the V class, which was noted for its increased hull strength, achieved by external framing between the inner and outer hulls. This allowed the submarine to dive to 45m/147ft 8in, as opposed to the norm of 30m/98ft 5in. Next in the pursuit of the ultimate diving machine came the *Nautilus*, later renamed *N1*, which was built over a four-year period at Vickers shipyards, went through various changes of design and cost £203,850. She was a sleek, fine-looking boat which, at 79m/259ft 2in in length, was at the time of her original design work almost twice the size of any existing submarine. The boat was regarded by many as a failure and never saw service of any description other than for training purposes and as a depot ship. However, that was indeed a short-sighted view. It was first and foremost an experimental boat, an early footprint on the road to the giants of the deep that were to follow. The next major development – outside of Germany's own fast and deep cruiser submarines, already discussed – came the one and only *X1*, which was another milestone in every way. She was a cruiser submarine of 3,600 tonnes/ 3,543 tons submerged – in no uncertain terms a true submersible cruiser, she could travel at 18.5 knots on the surface, was 110m/360ft 11in in length, and carried four 132mm/5.2in guns in twin turrets. Although conceived in 1915 – when Germany was already producing her cruiser boats – *X1* was not completed until 1925 and successfully passed a rigorous trials procedure. The trouble was, now there was no role for her in the British fleet, and the colossus was finally scrapped in 1937.

Other nations began to pursue the theme of giants of the deep. America produced its own *Nautilus* in 1930, following on from *Narwhal*, first introduced in 1927. Both were still effective and well used in World War II. They were massive boats – 113m/370ft 9in long, sleek in design and excellent to handle.

Elsewhere, other giants had emerged. The Japanese hired German technicians and began a programme of building submarines that went to the extremes of range and size, from midget subs to the largest cruisers, some capable of travelling 31,187km/20,000 miles and with an endurance of more than 100 days. The French also appreciated that size mattered and introduced the *Surcouf*, 110m/360ft 11in in length and 3,357 tonnes/3,304 tons, armed with two 203mm/ 8in guns and a light aircraft. Such was the fear and loathing of submarine warfare – loathing by the politicians because of the cost – that in 1930, Britain, America and Japan agreed to limit tonnage and impose armament limitations for all future cruisers, destroyers and submarines, with no new capital ships to be laid down until 1937. Neither France nor Italy agreed to sign, and in any event, the whole deal collapsed within three years because of developments in Germany.

**ABOVE: USS *Narwhal* (originally designated *V5*) was the lead ship in her class, commissioned in 1930, and was one of five submarines docked in Pearl Harbor when the Japanese attacked. Her gunners were quickly in action and shot down two of the torpedo bombers. BELOW: The one and only *X1*, the British cruiser submarine that led the field in the development of these underwater giants when built in 1925.**

**LEFT: USS *Barracuda*, (originally *V1*) was the first of the American V boats, commissioned in 1925, but at 2,032 tonnes/2,000 tons was considerably smaller than the giants that followed in the V-designation. She was recommissioned for service in 1942 to patrol the American Pacific coast. ABOVE: The pride of France, *Surcouf*, with a surface displacement of 3,302 tonnes/3,250 tons, was the largest submarine in the world when launched in 1929. She was lost with all 130 crew, supposedly sunk in the Gulf of Mexico in November 1942 amid accusations that she was spying for Hitler.**

# The inter-war years

While the maritime nations experimented with the *Nautilus* and similar boats, the main area of development in the 1930s was concentrated on building new workhorses for the various national submarine fleets. Britain had particular difficulty in achieving the aspirations of her admirals because of the country's parlous financial state, which was one of the reasons the government was so keen on maintaining international agreements curbing the arms race. However, in this regard, she was in danger of being left behind.

While Japan, Italy, France and Russia were now all focusing heavily on submarine manufacture, Britain had to rely on old stock well into the third decade of the century for the bulk of the service's work. The only new models apart from the experimental classes to appear beyond those ordered up during World War I were eight 1,524 tonne/1,500 ton O class and six slightly larger P class. They were excellent boats, designed for overseas patrols around the far-flung Empire, and especially the Far East. Nevertheless, they had a limited capability and also, by then, America, Japan, France and even Russia all possessed more submarines than Britain. Added to this was soon to be a new fear that would change the complexion of everything.

Adolf Hitler came to power in Germany, and secretly began to re-equip his military. Plans for the rebirth of the U-boat were already underway by 1934 when the Nazis issued bold statements repudiating the Treaty of Versailles, and in March 1935 demanded renegotiation of the permitted level of their armed forces. Britain's appeasement of Hitler in these times through the creation of the Anglo-German Naval Agreement allowed him to begin rebuilding the German Navy, supposedly to be kept within 35 per cent of the tonnage of the Royal Navy.

The single exception to this was in the dispensation for a new U-boat force, under the command of Admiral Karl Dönitz, which was allowed a tonnage 45 per cent of Britain's. The figures were both meaningless and unenforceable. Hitler made it clear that he would no longer be bound by a 20-year-old treaty and was already on the way to creating a new navy, although his plans for it were still far from complete when World War II began. Indeed, Dönitz himself had proclaimed in 1938 that it would take six years to rebuild Germany's U-boat force. In fact, as history now shows, the power house of the new Nazi machine achieved remarkable output and very soon the U-boats were back in business.

ABOVE: **Salute to the U-boats. These were Hitler's greatest sea-borne asset at the onset of World War II and throughout the conflict, given that Germany had been prevented from rebuilding the naval fleet that the nation was forced to surrender in 1918.** LEFT: **An early casualty in the undersea war: the British *Starfish*, commissioned as the third boat in the new S class in 1933, was depth-charged by German minesweeper *M7* in Heligoland Bight in January 1940. She was forced to surface and her crew went into captivity.**

ABOVE: **USS *Cachalot*, lead ship in the class bearing her name and another version of the V boats, lay in Pearl Harbor when the Japanese attacked. One of her men was wounded, but the submarine suffered no damage. Work on her was completed at a furious pace, and just one month later she sailed on her first war patrol and eight weeks later returned with vital intelligence of Japanese bases.**

Coincidence or not – but certainly by good fortune – the Royal Navy had begun its own modernization programme for the submarine service by ordering the building of a smaller boat specifically designed for operating in the North Sea and other confined theatres such as the Mediterranean. The boat to fill this role was another brilliant model, the S class, just 66m/216ft 6in long and displacing 886 tonnes/872 tons. In the fullness of time – and especially during World War II – the S class would prove to be one of the most important of all the Royal Navy submarines, with a total of 62 constructed over a 15-year period from 1935. This was to be quickly followed by the introduction of the T class, another exceptional boat particularly noted for the simplicity of its design. Approval for the build was given in 1935, and the first, *Triton*, came out for trials in 1938 to become the first of the 21 boats of this class built between 1937 and 1941. They were the largest class of ocean-going submarines to have been built by the Royal Navy, displacing 1,348 tonnes/1,327 tons surfaced and 83m/272ft in length, but only a handful were anywhere near ready by the outbreak of World War II.

The Americans, meanwhile, continued to be slow in their construction of new submarines. They were held back to some extent by their own financial depression, and also the fact that some in the US administration were of the same view of many naval traditionalists in Britain who still questioned the morality of submarine warfare and would have preferred the 1930 curtailment to have held.

However, with Hitler casting off the shackles of Versailles, Japan setting her sights across the Pacific and Mussolini already on a mission to build the New Roman Empire, the US Navy was forced to begin a rapid building programme, principally to meet the Japanese threat. By 1940, the American submarine force had answered its fundamental strategic questions and was rapidly enhancing its capability.

ABOVE: **The gathering storm. Hitler's submarine force was presented with a new flag when the swastika became the official emblem of the German nation in 1934.** BELOW: **An early workhorse of the US submarine fleet on America's entry into the war, USS *Sturgeon*, a 1,472-tonne/1,449-ton *Salmon*-class boat, was commissioned in 1938. She was operating in Hawaiian waters until December 1941 when she switched to war patrols in the Far East. One of the most notable events in her log occurred after her attack on a nine-ship Japanese convoy in 1943, when, in response, she was dived under a barrage of 196 depth charges and aerial bombs and survived to return safely home.**

# The Japanese story

TOP: **Japanese enthusiasm for submarine construction was already evident in the early 1920s, with *RO60*, which was one of nine L4 type medium series submarines, commissioned in 1923, with a surface displacement of 1,012 tonnes/ 996 tons.** ABOVE: **Early Japanese boats had many of the characteristics of U-boats, not unnaturally perhaps, since they took apart surrendered World War I German submarines and then hired many U-boat engineers and designers.**

Japan, more than any other nation, benefited from the era when Germany was prevented from reviving the U-boat force. Under the terms of international treaties to which the Japanese were signatories, their engineers – like those of Britain, American and France – were able to take apart surrendered U-boats and copy German technology. However, they went one better. They shrewdly hired more than 1,000 German technicians and designers who had built the U-boats in the first place – a move that no other nation had felt inclined to do. The Japanese government approved massive bounty payments to encourage the employment of German expertise, which even included former naval officers who had themselves commanded U-boats. They were offered five-year contracts, some with an annual salary of $12,000 plus bonuses, which was a huge amount of money at the time.

Armed with this double dose of technological know-how, Japan set about building a massive new submarine fleet. As early as 1928, American Intelligence reports noted with some alarm that Japanese efforts, based upon this German input was reaching worrying proportions. "Furthermore," one report stated, "as various submarines were completed, Japanese engineers gradually took over the entire production, from design to completion until finally a distinctly Japanese type of submarine gradually emerged." Indeed, by the mid-1930s Japan was in the final stages of creating the most versatile submarine fleet of any nation in World War II.

Even so, similarities to the U-boats were very evident in three main classes of submarine. The I class, for example, was the Japanese version of the fast and potent German cruiser submarine. It was the ultimate fleet boat, which Britain and America had also tried to perfect. It was also produced in several versions, including aircraft carriers and submarines that could carry multiple bombers; seaplane tenders; cargo ships; and some as submarine support vessels for the military. Next came the RO class, similar to but not as capable as the German coastal boats, and finally the HA class, which were smaller and intended for local defence and special missions, again sometimes attached to support ground forces. The most notable of the HA class, however, included midget submarines, one of the number of purpose-built boats developed in utmost secrecy by the Japanese in the late 1930s. They also built submarines large enough to transport the midget boats into

LEFT: **Another useful Japanese boat, the *HA101*, lead ship of the 376 tonne/370 ton SS type of small transport submarines, entered service late in the war but was well thought of, and closely examined when surrendered with the Japanese Navy's other surviving warships.** ABOVE: **Torpedo technology was also well advanced: these tubes housed the Type 95 torpedo that was based on the formidable Long Lance design, air launched and, some experts have claimed, was the best torpedo of the war.**

combat or enemy coastal areas where they would be released in swarms to attack – as it transpired – American warships. In fact, every aspect of Japanese technology throughout these years – obviously with an eye on the massive journeys involved in any future action in the boundless space of the Pacific – was geared to extremes in endurance and capability. They had the largest submarines in the world, the only boats with more than 5,080 tonnes/5,000 tons submerged displacement, or submarines over 121.9m/400ft in length – an achievement that was not matched until the arrival of nuclear-powered boats. They were said to have a range of 60,350km/37,500 miles at 14 knots, an achievement unmatched by any other diesel-electric submarine. These giants could carry three floatplane bombers, the only submarines ever built with such a capability.

Of the 56 submarines greater than 3,048 tonnes/3,000 tons built for action by the nations engaged in World War II, at least 50 were Japanese. They also led the field in diesel-electric submarines with 7457kW/10,000hp, and also in diesel-electric submarines capable of more than 23 knots surface speed – unmatched by any other nation.

Japanese boats also came equipped with some of the best torpedoes available. One significant development in this field was a torpedo known as the Long Lance, a submarine version of which, Mk95, was built with a 408kg/900lb warhead driven by an oxygen-fuelled turbine that left no wake, and was capable of travelling a full five miles at 40 knots. It was operative at the start of World War II and was well ahead of anything comparable built by the Allies at the time.

The Mk95 also carried the largest warhead of any submarine torpedo, initially 405kg/893lb, which was increased to 549kg/1,210lb in the final stages of the war. Another vital innovation in the Mk95 was the use of a simple contact

ABOVE: **The *I52* was a Type C-3 cargo boat, known as Japan's Golden Submarine because she was carrying £25 million in gold to Germany as payment for war material when she was sunk by a US Task Force. In 2006, attempts were made to raise the treasure by diver Paul Tidwell.** BELOW: **The sleek lines of the *I21*, a member of the elite Japanese Sixth Fleet, and the boat that famously shelled Australian cities in 1942, presumed lost with all hands in December 1943 while tracking an American convoy in the Pacific.**

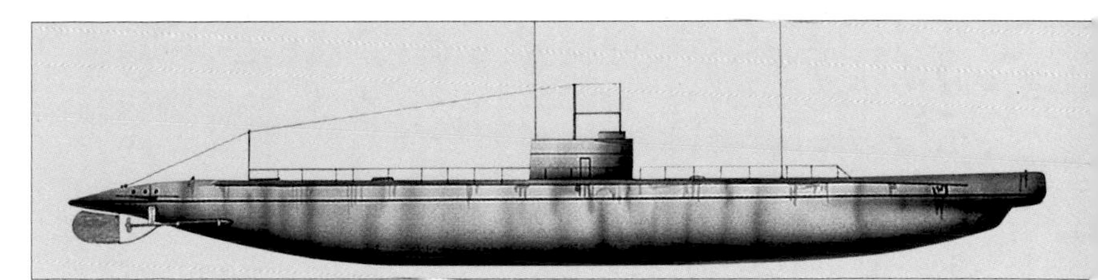

exploder, which was far more reliable than the American counterpart, the Mark 14, although the latter was improved early in 1944. Although there was no denying the sheer fire-power and versatility of the Japanese submarine fleet developed between the wars, and during the conflict itself, it has to be said that overall, Japan's submarine warfare – as will be seen – was not as effective as her masters expected.

LEFT: **A defining moment in submarine history, the arrival on the surface of the first men to be saved from a stricken boat using the experimental McCann Rescue Chamber, following the sinking of USS *Squalus,* which failed to surface after the final crash dive test in her last sea trials. She sank in 73.2m/240ft of water, and the surviving 33 crew waited in total darkness and without power for more than 24 hours before the rescue attempt began.** BELOW: **The *Squalus* badge, although when the boat was recovered, she was given a new name.**

# The sinking and recovery of USS *Squalus*

In the months leading up to World War II, two notable and tragic events occurred, both of which highlighted the perils faced by submariners of the participant navies in the coming months and years, and drew attention to the rescue possibilities from stricken boats. The two events, involving the US submarine *Squalus* and, across the Atlantic, the British boat *Thetis*, were remarkably similar in detail, but the outcomes were drastically different.

The *Squalus* left the Portsmouth Navy Yard, New Hampshire at 07:30 hours on May 23, 1939, for its 19th test dive under the command of Lt Oliver Naquin with 59 men on board – five officers, 51 enlisted men and three civilian inspectors. The exercise, part of a rigorous series of performance tests to which all boats were subjected, was to perform an emergency dive while cruising at 16 knots, diving to 15.24m/50ft within 60 seconds in order to avoid enemy attack. The dive was to be carried out just south-east of the Isles of Shoals where there was an average depth of 76.2m/250ft. When the order to dive was given at about 08:45 hours, the boat initially refused to respond, but within just a few minutes had settled gently on the bottom at 74.06m/243ft with the bow raised by 11 degrees.

The catastrophe had been caused by flooding of the aft section, and immediate action was taken to isolate the individual compartments to prevent total flooding, a procedure largely completed during the sinking. However, lighting had failed and only a few hand lanterns were available to relieve the darkness as the commander began the process of checking his situation. Of the 59 people who sailed that morning, 23 were in

the control room and 10 in the forward torpedo room. It was likely that everyone in the after battery room and both engine rooms had died. No contact was made with the after torpedo room. The possibility of survivors there remained, but communications with the control room had failed.

Saltwater was leaking into the forward battery and if it infiltrated the battery acid, chlorine gas might be released, or it could short-circuit the cells and cause a fire. This meant that the forward battery compartment, which was located between the two occupied spaces, would have to be left vacant. Meanwhile, a communications buoy attached to *Squalus* was released and surfaced soon after the sinking. Rockets were also fired from time to time and the sixth launched after four hours on the bottom was by chance

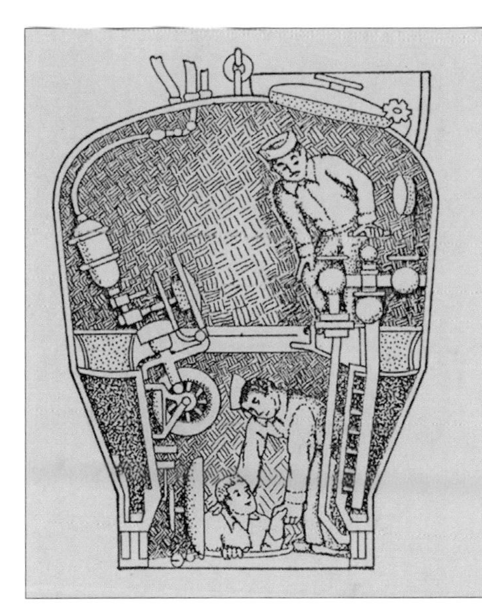

spotted by a lookout on the sister submarine USS *Sculpin*, which rushed immediately to the scene and located the communications buoy to establish contact. Just two minutes later, however, the connecting cable snapped. By mid-afternoon, two ships had arrived at the scene, one with an oscillator that enabled Morse code transmissions. As wives and families awaited news, a part-read message from *Squalus* transcribed as "condition satisfactory but cold" was taken to offer more hope than was the case.

In fact, by that time there were only 33 survivors, and their lives were at risk as the atmosphere in the boat reached twice the normal pressure. The US Navy's attempts to rescue the survivors went through various stages of discussion. The first, to pump out the flooded compartments to bring the *Squalus* to the surface was dismissed as too risky, because the reasons for the sinking were unknown.

Finally, on the morning of May 24, the submarine rescue vessel USS *Falcon* arrived above the submarine with a team of divers and a McCann Rescue Chamber to begin a dramatic rescue attempt that was to last a full ten hours. The divers were commanded by Lt Commander Charles Momsen, who had invented both the Momsen Lung and the diving bell. The latter would be attached to an escape hatch over the forward torpedo room allowing survivors to be drawn to the surface by cable. The theory was fine, and well rehearsed in training conditions, but in practice this was to be an historic, drama-filled mission. By 10:07 hours, divers had managed to get the diving bell over the escape hatch and the crew inside the *Squalus* banged out a welcome on the hull. However, this was merely the beginning of a long and exhausting process, as the bell with its two operators for the motors, ballast, air pressure and communications sought to make the connection to begin the first evacuation, a task that overall would take two hours or more from leaving the mother ship to returning with survivors, with a maximum of eight men at a time. All went smoothly until the last trip, the one carrying Commander Naquin and the last seven of his crew. They were safely aboard the diving bell by 19:50 hours to begin the ascent when the cable attached to the *Falcon* fouled and frayed. It took four and a half hours to bring the chamber to the surface. Nevertheless, all 33 survivors of the original accident were safe.

The boat was salvaged in the summer of 1939, and after repair was recommissioned as USS *Sailfish*. She went on to survive World War II, sinking 40,642 tonnes/40,000 tons of enemy shipping.

ABOVE LEFT: **The internal lay-out of the McCann diving bell, which could extract crew through the escape hatch and return to the surface carrying up to eight passengers on each trip.** ABOVE: **During a three-month salvage effort by the US Navy using air tanks on the surface, the bodies of those who died were removed and the *Squalus* finally returned to the surface.** LEFT: **The salvaged submarine in a sorry state. She was immediately sent to the shipyards for repairs and was recommissioned to serve for the duration of World War II under her new name, USS *Sailfish*.**

# The tragic loss of the *Thetis*

Barely a week after the *Squalus* tragedy, an accidental sinking bearing incredible similarities occurred off the coast of Wales where a brand-new British submarine went to the bottom, with totally catastrophic results. It was indeed a disastrous event for the Royal Navy with so much riding on the new T-class submarines just weeks before the outbreak of World War II. HMS *Thetis* was the third of the T boats scheduled for commissioning, and after months of rigorous tests since her launch in 1938, she was to begin final acceptance trials in late May 1939. Her captain, Lt Commander Guy Bolus, had stood by her during her fitting out and, with a crew of five other officers and 48 ratings, left the yard at Merseyside for trial runs in the Clyde estuary during May. There were a few problems to be ironed out, which was not unusual with a new boat, and *Thetis* returned to Cammell Laird yards for adjustments. The final diving trial was postponed until June 1, in Liverpool Bay.

*Thetis* sailed from Birkenhead soon after 09:00 hours that day with an unusually large number of people on board. In addition to the crew, there were officers from the flotilla *Thetis* was to join when she was fully commissioned, civilian observers from the Admiralty, 26 employees of shipbuilders Cammell Laird, five from other shipbuilding firms who would be building T-class boats, two members of staff from City Caterers of Liverpool, who were supplying the luncheon, and a Mersey pilot: 103 men in all. The weather was fine, the sea was calm with only a touch of an easterly wind. Everyone on board was looking forward to an interesting trip, and for the non-sailors it was an adventure to be enjoyed with perhaps a touch of apprehension. Accompanying *Thetis* during the initial stages of the trial was the *Grebecock*, a two-year-old steam tug manned by a crew of seven and laid on by Cammell Laird as part of their contract with the navy. On board was a navy observer and submariner, Lt Dick Coltart, whose task it was to keep a general watch and to warn any ships that might sail into the diving area. The *Grebecock* was also to stand by to take off any civilian passengers from *Thetis* who might wish to disembark after lunch, but none chose to do so. Then came a signal from *Thetis* that she would be diving for three hours. Coltart watched as Lt Commander Bolus left the bridge and the conning tower hatch was closed. At precisely 14:00 hours, Bolus gave the order to dive. But *Thetis* stubbornly refused to budge. Minutes later, she disappeared beneath the waves very quickly, and it was assumed that all was now going to plan.

Unfortunately it was not. The 533mm/21in wide No. 5 torpedo tube was open to the sea and the internal rear door of No. 5 tube had also been opened because a faulty test drain pipe showed empty when, in fact, the tube was full. Therefore, the door was open and *Thetis* sank. The situation was critical. As time passed, Coltart became convinced the submarine was in trouble and sent a carefully worded signal to the submarine base at Fort Blockhouse. Unfortunately, the telegraph delivery boy's bike got a puncture en route and the message was

LEFT: **HMS *Thetis*, one of the new British T boats, launched into her acceptance trials in May 1939 in the Clyde estuary before finally beginning diving trials in Liverpool Bay on June 1. On that day, she had a full crew of 54 officers and ratings aboard, along with shipbuilders, civilian observers and other guests, 103 men in all, when disaster struck.**
ABOVE: **The *Thetis* badge that was never operationally used.**

ABOVE LEFT: **Pristine again after the dramas below, the *Thetis* was recovered and recommissioned as *Thunderbolt* to enter service in World War II.** ABOVE RIGHT: **As *Thetis*, she went upside down and stuck firm in the mud when on her pre-acceptance trails. Although rescue ships reached her after some delay, they could not gain access before she finally sank to the bottom with great loss of life.** LEFT: **The salvage boats hover as the recovery begins.** BELOW: **The *Thetis* is brought to the surface and dragged ashore for repairs and renewal.**

delayed. When it did arrive, fast destroyers were ordered to the scene and the RAF sent a flight of four Ansons, conducting low-level fly-bys until dusk closed in. A salvage vessel, *Valiant* sailed from Liverpool but by then it was almost eight hours since any word had been heard from *Thetis*.

Three hours later, searching over a wider arc, a crewman on the destroyer *Brazen* spotted something in the distance, and as the ship headed towards it, an incredible sight became visible – 5.5m/18ft of the stern of the 83.5m/274ft-long *Thetis* was protruding from the sea. She lay at an angle, nose down in 45m/147ft 8in of water 24km/15 miles off the North Wales coast. Inside the boat, conditions were desperate. The air was foul and rapidly running out. Bolus decided that the only solution was to make an attempt to leave the boat, using the David Submerged Escape Apparatus (DSEA), although few men were fully trained in the procedure. Volunteers stepped forward. Captain Joe Oram, officer commanding the 5th Flotilla, and Lt Woods were selected and successfully made their escape, popping to the surface within seconds. They were picked up by *Brazen* and proceeded to give details of the situation below.

Four more men escaped later, but after that, no further exit attempts were made, for unknown reasons. By then, *Brazen* had been joined by six destroyers, five tugs, the Llandudno lifeboat and the salvage vessel *Vigilant*, all strewn in an arc around the stricken submarine.

One final hope remained: to cut a hole in the side of *Thetis* with oxyacetylene gear, for which that section of the boat had to be raised above the surface. At 15:10 hours on June 2, as efforts were being made to achieve this, the wire rope snapped and *Thetis* plunged back to the bottom, taken all remaining 99 personnel with her. After further attempts to reach the boat failed, the rescue efforts were abandoned the following day amid a critical uproar from relatives. An inquiry did nothing to satisfy the relatives and trade unions, and a long and bitter legal action ended six and a half years later in the House of Lords, where claims for negligence were finally rejected on the grounds that there was no proof of liability.

*Thetis* was raised three months after she sank. Human remains still on board were retrieved for burial. The boat was repaired and went back to sea under the name of *Thunderbolt*, and had a successful campaign in the Mediterranean before being depth-charged north of Sicily on March 14, 1943, sinking for the second and last time.

# The action begins, and *Athenia* goes down

Just after 11:15 hours on September 3, 1939, Royal Navy ships and submarines received the signal: "Commence hostilities with German forthwith." Four minutes later, the submarine *Spearfish* commanded by Lt J.H. Eden delivered a torpedo towards a German U-boat in the North Sea. It missed the target but established the first confrontation with the enemy. The Nazi High Command gave similar instructions to its U-boats that afternoon and before the day was out, Germany had launched into a controversy that matched its World War I attack on the passenger liner, *Lusitania*. This time, the victim was the 13,717 tonne/13,500-ton passenger liner, *Athenia* carrying 1,103 civilians, including more than 300 Americans attempting to escape the war in Europe. She became the first British ship sunk by a German U-boat in World War II.

The *Athenia* had already sailed from Glasgow en route to Montreal when war was official declared, and that afternoon as she steamed 402km/250 miles north-west of Inishtrahull, Northern Ireland, she was spotted by *U30* under the command of Lt Fritz-Julius Lemp. Regardless of the fact that German U-boats were supposed to be operating under prize regulations that obliged them to stop and search any potential targets, Lemp ordered his crew to open fire without giving any prior warning.

In the controversy that followed, and which remained an issue to be considered in the Nuremberg War Crimes trials, it was passed off as a case of mistaken identity. Lemp stated that he believed the *Athenia* was an armed merchant cruiser, and fired two torpedoes. The ship sank quickly with the loss of 112 passengers and crew, including 28 Americans, although that did not prevent President Franklin D. Roosevelt continuing with his plans to declare America's neutrality in the war. Some of the passengers claimed they were attacked again when they were at sea, escaping in lifeboats.

TOP: **A photograph of the *Athenia* in the final throes of her sinking, apparently taken from the one of the rescue boats that rushed to the scene after the alert of the attack upon her by the *U30*. Three British destroyers and three merchant ships aided the rescue, which saved 90 per cent of the 1,103 civilians aboard.**
ABOVE: **The majestic liner in all her glory, a ship that the U-boat commander who ordered the sinking claimed he mistook for an armed merchant cruiser. Few believed him.**

Lemp later offered an explanation. He stated that having received three radio messages sent to all Kriegsmarine vessels stating that war existed between Britain and Germany, he spotted the ship heading away from Britain in a north-westerly direction. Having ordered his boat to dive to periscope depth, he

FAR LEFT: **Fritz-Julius Lemp, captain of *U30* in the controversial sinking of the *Athenia*. He later sank a total of 17 ships in the same submarine, becoming the seventh U-boat commander to be awarded the Knights Cross at the age of 26. He met his match in May 1941: he was killed in action when the British captured his new submarine, *U110*, with its vital Enigma machine.**

ABOVE: **The killer blow to the *Athenia* may well have been struck by this torpedo, one of 11 that the Type VIIA *U30* could stow on each mission, a task made easier on July 7, 1940, when the boat became the first German submarine to make use of the newly captured French bases, docking in Lorient.**

BELOW: ***U30*, one of the busiest in early exchanges of the war and, as the front boat in Germany's 2nd Flotilla, completed eight patrols before she became a training submarine.** BOTTOM: **The British submarine *Oxley* lost off the coast of Norway after being mistaken for a U-boat and torpedoed by companion submarine *Triton* in September 1939.**

saw that the ship was blacked out and came to the conclusion that she was an armed merchant cruiser, and therefore a legitimate target. He fired two torpedoes, the first striking the ship squarely, and the other missing. He claimed he dived to avoid the possibility of being hit by the second torpedo and, returning to periscope depth, fired a third because the ship did not seem to be sinking. This too missed, but undoubtedly ran towards the escaping passengers. At this point, Lemp checked his Lloyd's Register and discovered his mistake.

Soon afterwards, *U30* intercepted a plain-language transmission from the stricken ship identifying itself as the *Athenia*, but even then he did not attempt to aid survivors, as Prize Rules required, and maintained radio silence throughout. Thus, in an instant the memories of Germany's U-boat campaign of unrestricted warfare in World War I were revived.

In fact, the effect worldwide was such that Hitler himself decreed that all such accusations would be dealt with by categorical denial and Goebels even put out a statement accusing Britain of deliberately sinking the *Athenia* to stop Roosevelt declaring neutrality. The Nazis even went to the extent of falsifying records to show that *U30* was at least 322km/200 miles from the *Athenia* at the time. The truth, however, was confirmed at Nuremberg by Adolf Schmidt, a member of Lemp's crew who had been wounded in a later action and was left in Iceland, where he became a prisoner of war. Even so, the international outcry did have immediate repercussions when Hitler signed an order to prevent further attacks on passenger ships, regardless of the nation of origin.

The British, meanwhile, also proved that accidents could happen in this electrifying turmoil at the start of the war. On September 10, an old O-class submarine, *Oxley*, set off on her first major patrol off the coast of Norway. In heavy seas, at night and in poor visibility, she was spotted in the distance by one of her companion submarines, *Triton*. She failed to make the correct recognition signal. As a result, she was mistaken for a U-boat and torpedoed, and became the first submarine loss of World War II. Of the 54 crew, only her captain, Lt Commander H.G. Bowerman, and one other survived.

LEFT: **The explosion caused by depth charges is violent and dramatic, and Lt Sheldon H. Kinney, Commander, USS *Bronstein* described them as "having a magnificent laxative effect on a submariner". They were used by ships or aircraft to attack submerged submarines and were first developed by the British for use against German submarines in World War I but were substantially improved by 1939.** BELOW: **Firing depth charges from the deck of a ship.**

# Mines, depth charges and leaky boats

Heligoland Bight and Skagerrak once again became the early scene of battles between British submarines and their German opponents, and Sir Max Horton, hero of great actions in World War I and whom the Germans had called a "pirate", was now directing operations under the personal eye of his good friend and First Lord of the Admiralty, Winston Churchill. As the battles began in earnest, they identified a particular target for the submarines in the unhindered supply of iron ore, which was being turned into ships and aircraft, coming down into Germany from northern Sweden – exactly the same problem that first emerged in 1914. The ore was being trans-shipped through the ice-free port of Narvik in northern Norway, still a neutral country.

On the same day, the British War Cabinet pronounced that British submarines could and would henceforth sink transport ships in the region on sight, and two days later it extended that order to cover any ship sighted within 16km/10 miles of the Norwegian coast. Max Horton had committed virtually his entire force to the Norwegian campaign a week earlier and although the British force would stay well away from the Baltic, where Horton had made his name, he dispatched every available submarine to join those already on patrol to cover all the German exits from Heligoland Bight, Skagerrak and the Kattegat.

The Skagerrak strait, between the southern coast of Norway and the Jutland peninsula of Denmark, was a vital seaway. Together with the Kattegat strait, it connects the North Sea with the Baltic Sea. About 214km/150 miles long and from 121km/75 to 145km/90 miles wide, the Skagerrak is shallow

near Jutland but deepens near the Norwegian coast. It was in these treacherous waters that Max Horton placed an arc of submarines at the beginning of 1940. These were supplemented by others laying 50 mines apiece on each trip, and in the first half of the year there was probably not a single home-based submarine commander who did not see service there. Several would never return.

The extent of their activity is demonstrated by the amount of shipping sunk by the submarine service in the month between April 8 and May 4: close on 35 assorted vessels, of almost 90,000 tonnes/88,579 tons, with the loss of four British submarines. It was a dangerous time, particularly for the less experienced submariners, because it was the first time in history that efficient depth charges were used in abundance, in addition to the normal hazard of mines.

ABOVE LEFT: **Destroyer Escorts accompanying shipping movements routinely carried a heavy load of 100 or more depth charges to ward off enemy submarines or preferably kill them off completely. To be fully effective, they had to be fired with extreme accuracy.** ABOVE RIGHT: **An even greater threat to both surface ships and submarines alike was the greatly improved technology for creating minefields. The toll was heavy on both sides.** LEFT: **The British submarine minelayer, *Porpoise*, who made her name helping to break the blockade of Malta with food and fuel stacked on her mine deck. She subsequently fell victim to either mines or depth charges while on minelaying patrols in the Far East.** BELOW: **The jubilant crew of the British boat *Truant*, which was forced to remain on the bottom for a long period to escape the retribution of depth charges after torpedoing the German cruiser *Karlsruhe*.**

Depth charges first came into use during World War I. The early forms of the weapon consisted of large cylinders containing TNT, which were rolled or catapulted from the stern of a ship. In the intervening years, depth charges and machinery for launching them had improved dramatically, and experimental work on new designs was ongoing and came into use in the early stages of World War II. The use of a new and more powerful explosive called Torpex, a mixture of RDX, TNT and aluminium, substantially reduced the size of depth charges. The casings were also streamlined to make them sink faster and thus more difficult to avoid. To have literally dozens of these sinking towards your boat and exploding all around as the commander looked worryingly at his depth gauge was one of the most frightening experiences imaginable.

The T boat *Tetrarch*, under the command of Captain Roland Mills, took the record for staying submerged for the longest known period to date of 43 hours while under heavy counter-attack by depth charges. When he finally came up for air, his crew were literally falling as if drunk. He confessed in a memoir: "The effects of successive long spells under water, accompanied by depth-charge attacks, could have untoward effects, especially when the charges were poured into the sea by a posse of marauding destroyers, angry because the submarine had just fatally damaged one of Germany's brand-new cruisers."

That is exactly what happened to Admiral C.H. Hutchinson, then a Lt Commander of another of the early T boats, *Truant.* He had carried out a number of successful patrols in the North Sea during the Norwegian campaign, when in April 1940 he, along with other boats, was to provide protection and diversion for what proved to be the ill-fated Allied landings of troops at Narvik to counter the German invasion. One evening his attention was drawn to propeller noises, very faint, to the north. The sounds gradually got louder, and eventually he saw through the periscope a German cruiser dead ahead, screened by an escort of four small destroyers. He fired a large salvo of torpedoes and raised his periscope to take a look. He could see nothing but a very enlarged after-part of a destroyer which was right on top of him.

Unbeknown to Hutchinson at the time, he had torpedoed the famous German cruiser *Karlsruhe,* which was homeward bound from Oslo. The Germans were naturally furious. The cruiser was abandoned and was so badly damaged that Berlin ordered that she should be scuttled immediately.

# Heroic work in the Mediterranean

The Norwegian campaign was followed quickly by the Battle of Britain, and the Royal Navy and the submarine force played its own part defending the potentially vulnerable eastern sea board as the threat of invasion mounted in the summer of 1940. It also provided cover for the massive operation to rescue the British Expeditionary force from France, while at the same time confronting the rising tide of U-boat activity in the Atlantic, soon to be facilitated by the use of France's Atlantic ports. But even as the demands of this new menace grew, a fresh area of conflict at sea opened up in the Mediterranean when Mussolini joined forces with Hitler and decided to rebuild the Roman Empire out of British possessions in North Africa and the Suez Canal in particular. One of the keys to this operation was his intention to capture the island of Malta, where defences were woefully meagre.

From the beginning, British resources were stretched to the limit, especially for the submarine flotillas, such as they were. Overnight, traditional British supply routes to its eastern outposts were closed down. The British had to send their ships on the long journey, lasting two or three months, around the Cape of Good Hope for all supplies, other than those borne by special convoys with major fleet escort operations, which the Royal Navy could ill afford. With the lack of very serious competition, the Italian Navy put up a very worthwhile case for supremacy in the Mediterranean, fielding five capital ships, 25 cruisers, 90 destroyers and 90 excellent submarines under a very capable command structure. They had back-up from 2,000 front-line aircraft and were undoubtedly leading the West in the diversity of experiments with mini-subs, human torpedoes, underwater chariots for frogmen and exploding motorboats. When Mussolini entered the war, the British had no real naval force between Gibraltar and the Italian fleet ports. The British base at Malta possessed a puny collection of just six submarines. Three of them, the minelayer *Grampus* and *Odin* and *Orpheus*, were lost in the first week after Mussolini

**ABOVE: British submarine *Upholder* sank or damaged 22 enemy ships, which earned her skipper, Lt Commander Malcolm Wanklyn, the award of a Victoria Cross, though sadly he and his boat did not survive the war. INSET: The badge of HMS *Upholder* linked by the crest as a tribute to the original Wanklyn boat when a new British diesel electric submarine was launched in 1987. BELOW: Submarines of the 10th Flotilla based at Lazaretto, Malta, went into maritime history for their exploits. They included *Upright* (second from right) which sank Mussolini's cruiser *Armando Diaz*.**

signed his pact with Hitler. It was believed that they were all sunk with no survivors by deeply laid Italian mines.

Very soon, however, the British submariners began to show their mettle. Substantial totals of enemy shipping were sunk, and none better than that achieved by the brand new *Upholder,* under the command of Lt Commander Malcolm Wanklyn. He became the most successful submarine commander in World War II in spite of what would prove to be a relatively short career. *Upholder* arrived for duty in the Mediterranean on January 12, 1941. On her first patrol, she sank a 5,000-tonne/4,921-ton cargo vessel in one of the Axis convoys just off Tripoli, and in almost the same spot two days later sank an 8,000-tonne/7,874-ton supply ship. This time, *Upholder* was counter-attacked and took the brunt of 32 depth charges before escaping. After a brief lay-off, *Upholder* returned to the fray and in April sank three separate targets in a single convoy: a destroyer, a 4,000-tonne/3,937-ton supply ship and a 10,000-tonne/9.842-ton supply ship. On May 24, Wanklyn sighted a troopship, the Italian liner *Conte Rosso* of 17,800 tonnes/17,519 tons, carrying 4,000 troops, with an escort of three cruisers and several destroyers. He launched two torpedoes, which both scored direct hits amidships. The liner sank rapidly, killing 1,500 of the men on board. Wanklyn was awarded the DSO for this attack.

However, it was not all good news. During those vital months, the British submarine force in the Mediterranean lost four of its own boats in quick succession. *Usk* struck a mine at Cape Bon, *Undaunted* and *Union* were both depth-charged off Tripoli and the minelaying submarine *Cachalot* was rammed by an Italian torpedo boat. Malcolm Wanklyn and his fellow commanders began to take greater risks and shorter breaks. In September 1941, as Wanklyn set out on his fourteenth patrol, he joined up with *Unbeaten, Upright* and *Ursula* for a prearranged joint-assault to intercept a troop-carrying convoy leaving the southern Italian port of Taranto bound for Tunisia.

ABOVE: **The minelayer *Grampus*, lead boat of her class, built in 1935, was deployed off the coast of Sicily in June 1940, when she was spotted by a group of Italian torpedo boats and torpedoed. There were no survivors.** BELOW: **The *Undine*-class submarine *Ursula*, also in the Mediterranean flotilla, was noted earlier in action near the *Elbe*, when her captain, Lt Commander G.C. Phillips, courageously dived beneath a screen of six destroyers to sink the German cruiser *Leipzig* at a range so close that the *Ursula* herself was badly shaken by the explosion.**

Two torpedoes from *Upholder* hit the transport *Neptunia,* which sank in eight minutes. Wanklyn's third torpedo hit *Oceania* and ripped away her stern. The third transport, *Vulcania,* was hit by *Ursula* from long range, and *Utmost* finished her off as she tried to limp away. Meanwhile, *Utmost* now sighted three Italian cruisers and several destroyers and promptly sank one of the cruisers. As a result of his latest exploits, Malcolm Wanklyn was awarded the Victoria Cross in December 1941. On April 6, 1942, *Upholder* sailed out of the base for her twenty-fifth and last patrol before heading home for a refit, but she never made it back. On April 14, *Upholder* was depth-charged and sunk by the Italian torpedo boat *Pegaso* north-east of Tripoli. Wanklyn's was a remarkable record. In 16 months, *Upholder* had sunk or damaged 22 enemy ships, including three U-boats, two destroyers and one cruiser as well as 120,910 tonnes/119,000 tons of supply ships.

# "Finest work" by British submarines

Much action remained, and more VCs were to be won in the continuing saga of the battle for the Mediterranean. Among them was an incident involving *Thrasher.* She had been in the Mediterranean only a couple of months and in February 1942 was operating off northern Crete. There, on February 16, she penetrated an escort screen of five destroyers in broad daylight and torpedoed and sank a large Axis supply ship off Suda Bay. The destroyers gave chase, and as she crash-dived they remained overhead for an hour or more, sending down more than 30 depth charges. She surfaced after dark and headed off towards the Gulf of Taranto, but very soon banging noises were heard on the casing. Two unexploded bombs had become lodged on the submarine.

Lt Peter Roberts and Petty Officer Tommy Gould volunteered to go out and recover the bombs. The first bomb was located and gently removed, wrapped in sacking and dropped over the stern. The second was lodged in a narrow space and they had to lay full length in complete darkness, pushing and dragging the bomb for a distance of some 6.1m/20ft until it could be lowered over the side, a task which took more than half and hour. There was a very great chance that the submarine might have to crash-dive while they were under the casing. Had this happened, they would have been drowned. Robert and Gould deservedly won the VC for their heroic action.

Throughout this period, there were many acts of great gallantry and daring among the submarine flotillas in the Mediterranean, many of which resulted in their final demise. In March 1942, the *Torbay* won particular attention and a Victoria Cross for her captain, Lt Commander Anthony Miers. He sighted a north-bound convoy of four ships too far distant for him to attack initially, and he decided to follow in the hope of catching them in Corfu harbour. During the night, *Torbay* approached undetected up the channel and remained on the surface charging her battery. Unfortunately, the convoy passed straight through the channel, but on the morning of March 5, in glassy sea conditions, Miers successfully attacked two store ships and then brought *Torbay* safely back to the open sea. The submarine endured 40 depth charges and had been in closely patrolled enemy waters for 17 hours.

*Turbulent* also had numerous successes recorded against her name and represented a remarkable achievement which eventually brought the award of a Victoria Cross for the "great valour" of her captain, Commander John Wallace "Tubby" Linton, at the end of what was ultimately a tragic journey. In March 1943, Linton had already been hailed as one of the greats in the submarine hall of fame, but his was not simply a story of "big hits" in the Mediterranean. It went well beyond that, in terms

ABOVE: **A classic shot of the periscope in operation in the control room, and without which the submarine would be useless. The skills of the operators in judging the distance, heading and speed of the target were imperative when firing a torpedo.**

of courage, daring and dedication to duty in the face of great adversity, as recorded in the citation for the VC he was awarded: "From the outbreak of war until *Turbulent's* last patrol, Commander Linton was constantly in command of submarines, and during that time inflicted great damage on the enemy."

The journey for Linton and his crew was drawing to a close in February 1943, when they sailed from Algiers for a patrol in the Tyrrhenian Sea. On March 1, she attacked and sank the Italian steamship *Vincenz*. On March 11, she torpedoed the mail and supply ship *Mafalda*. She was herself attacked the following morning by the anti-submarine trawler *Teti II* but escaped, only to hit a mine off Maddalena, Sardinia, on March 23. The discovery of her wreck many years later confirmed the irony of her last voyage.

The part played by the submarines of the 1st and 10th flotillas has not always been fully recognized by historians, although Commander F.W. Lipscomb was moved to record: "There is no doubt that the whole Mediterranean offensive can be regarded as the finest work ever accomplished by British submarines." In the crucial two years between January 1941 and December 1942, the Italians lost 171 ships in the Mediterranean, totalling well over half a million tonnes/492,103 tons, a large proportion of which were sunk by the submarine flotillas. On the debit side at that time, Britain lost 14 submarines in the Mediterranean, with 70 officers and 720 ratings.

TOP LEFT: **HMS *Thunderbolt*, the salvaged and recommissioned *Thetis* which re-emerged in October 1940, later adapted to carry Britain's "human torpedoes", known as Chariots, before she was lost again, this time sunk by an Italian corvette in March 1943.** TOP RIGHT: ***Sokol*, of the Polish Navy, in Malta Harbour, flying the Jolly Roger to denote her successes. She was previously HMS *Urchin*, loaned to the Poles after the loss of their own submarines.** ABOVE LEFT: **Survivors of a U-boat rescued by Allied vessels in 1943, the year in which the Germans saw their highest number of losses, with 242 boats sunk.** ABOVE RIGHT: **The uncommon sight of a U-boat in the Mediterranean: only 62 German boats passed through the Straits of Gibraltar throughout the war and none of them returned: all being sunk, severely damaged or scuttled.** BELOW: **A memorial tribute in Valetta to the British and Allied submarine forces for their efforts during the war.**

TO COMMEMORATE THE CLOSE TIES FORGED BETWEEN THE PEOPLE OF MALTA G.C. AND THE OFFICERS AND MEN OF THE BRITISH AND ALLIED SUBMARINES BASED IN MALTA IN H.M.S. TALBOT, LAZARETTO MANOEL ISLAND IN TIME OF WAR, 1939-1945

# U-boat war in the Atlantic

In World War I, U-boats operating in the Atlantic almost brought Britain to her knees. The same procedure was attempted by Admiral Karl Dönitz, head of the German submarine service, by attacking Allied shipping at its weakest point – merchant vessels – with what was, at the time, a comparatively small fleet of U-boats. The aim was to harass the movement of essential British troop convoys to the Mediterranean and beyond, and cut Britain's vital seaborne supply routes. This developed into what Churchill christened the Battle of the Atlantic.

Restrictions placed on Germany's re-armament during the inter-war years meant that Hitler began his war with just 46 operational vessels, of which just over half were submarines. Dönitz ensured this state of affairs was quickly rectified as German shipyards began turning out the new-era U-boats at an incredible rate. Notwithstanding the *Athenia* debacle, the U-boats opened their account against the British with spectacular results very early in the war.

*U29*, under the command of Kapitän-Leutnant Otto Schuhart, sank HMS *Courageous*, one of Britain's four aircraft carriers, off the coast of Ireland on September 17, 1939. The ship went down in 20 minutes, taking with her 518 of her 1,200 crew. On October 13, a U-boat adventurer, Kapitän-Leutnant Günther Prien, commanding *U47*, set course for Scapa Flow, the Royal Navy's northern anchorage, hoping to do some damage. The fleet was out, but the fine old battleship *Royal Oak* was standing guard and Prien promptly dispatched her to the deep with his torpedoes. In all, 833 men lost their lives, and many of the 396 survivors suffered severe burns and life-threatening injuries.

**TOP: A rare shot of battlecruiser *Gneisenau* and *U47* together. The latter, having entered the supposedly impregnable Royal Navy base at Scapa Flow in October 1939, sank the battleship *Royal Oak*, with the loss of 833 men. *Gneisenau*, operating with her sister ship *Scharnhorst*, infamously sank the carrier *Glorious* and two escort destroyers, leaving 3,000 men floundering in the ice-cold waters of the North Atlantic. Only 36 survived. ABOVE: A view of an Atlantic convoy of merchantmen from one of the many destroyers that screened their journeys across the Atlantic carrying vital supplies.**

By repeating their tactics of World War I, the Germans – having failed to invade – hoped to subdue the British by starving the nation of food and equipment. Germany's main naval weapon was to be the U-boat, and control of the French Biscay ports provided bases from which they could sally forth into the Atlantic without having to pass either through the Channel or around the north of the British Isles at the end of every patrol. The convoy system of bringing in supplies was operated from the outset and in the first full two years of the

FAR LEFT: **The popular wartime poster expressing the heartfelt feelings of many: "Let 'em have it!" Similar poster art was also rife in Germany.** LEFT: **Admiral Karl Dönitz, naval strategist, was a mastermind of the U-boat campaign and briefly Hitler's successor after the Führer committed suicide.** ABOVE: **The Kriegsmarine U-boat golden badge of honour.**

BELOW: **U-boat wolf-packs made a direct course for the American Liberty ships in the Atlantic convoys. More than 2,750 such ships were built in the US to mass-produced design, each in about 70 days at a cost of less than $2 million per vessel. They could cross the Atlantic at an average speed of 11 knots, carrying over 9,144 tonnes/9,000 tons of cargo in the five holds, plus airplanes, tanks and locomotives lashed to the deck. A Liberty could carry 2,840 jeeps, 440 tanks, or 230 million rounds of rifle ammunition.**

war, 12,057 ships arrived at British ports in 900 convoys. The convoys grew in size: eventually between 40 and 60 ships would steam in columns with 3.2km/2 miles between each column and 0.53km/0.33 miles between each ship. A 12-column convoy would extend almost 11.1km/6 nautical miles in length and almost two miles deep. They needed massive protection: escorts of surface ships, destroyers, submarines and aircraft, when in range, would accompany convoys at various stages en route.

However, the U-boats gradually built their devastating campaign against Allied ships, supplemented by mines, aircraft and surface ships, and by the end of 1940 alone, 3,048,141 tonnes/3,000,000 tons of Allied shipping were lost, and the tonnage escalated dramatically when Dönitz introduced the infamous wolf-pack tactics and night surface attacks. The admiral often directed wolf-pack operations himself. A group of U-boats would patrol in lines scouting for convoys and once spotted, one boat would act as the shadow, reporting its heading and speed to U-boat headquarters. Other boats would be directed to form around the convoy, and at a given time, when all were gathered and in position, launch a combined attack, usually on the surface at night, thus greatly reducing the effectiveness of ASDIC.

A number of German U-boat commanders made their name as wolf-pack specialists, notably Günther Prien, whose *U47* sank 31 ships and damaged eight others before he and his crew were lost when their boat went down in battle on March 7, 1941. Prien, then just 33 years old, was one of the most highly decorated U-boat commanders, and the first to be awarded the Knight's Cross – a feat mentioned in the post-war writings of Winston Churchill. Another famous wolf-pack expert was Otto Kretschmer, who was 28 years old in 1940 and in command of *U99*. He sank almost 304,814 tonnes/300,000 tons of shipping and, like Prien, was awarded the Knight's Cross with Oak Leaves, and later the Knight's Cross with Oak Leaves and Swords. He survived the war and lived to the age of 86.

Almost 140 U-boat wolf-packs were assembled during 1940–43. Their operations were meticulously recorded, some fast and effective, others lasting ten days or more. The number of U-boats in each pack ranged from a minimum of three to around 20 in the biggest groups. This phase of the war became known among the Germans as the "happy time", a constant wearing down of merchant shipping that, in fact, became a far greater threat than the possibility of invasion. Eventually, however, brilliant work by scientists and all branches of the armed forces combined to bring an end to this devastating instrument of attrition.

LEFT: **USS *Pompano*, one of the *Perch*-class of American submarines, built in the mid-1930s, which saw great service against the Japanese. She was also one of the many losses, having left Midway in August 1943 bound for the coasts of Hokkaido and Honshu. She sank two ships in early September, but was never heard from again, presumed lost to a mine.**
ABOVE: **Poster art in America, blunt and to the point.**

# Pacific encounters of America and Japan

American naval experts are the first to admit that prior to their nation's entry into World War II in December 1941, their submarine capability was faulty and neglected. This was due in part to the numerous international treaties that the US had supported on the banning of unrestricted submarine warfare, and due also to the general malaise about getting involved in the war. There were other vital flaws in America's somewhat archaic peacetime submarine strategy in that many considered it to be an unduly hazardous occupation, and suicidal if operating within range of an enemy air base. Thus, the Silent Service amounted to little more than one per cent of the total resources of the US Navy at the beginning of the 1940s. Also on the debit side was the fact that American torpedoes were not always effective, a fact which was not fully corrected until well into the war.

The U-boat campaigns in the Atlantic changed many minds: suicidal or not, an effective submarine force was a necessary evil to be embraced by the US, a fact which was finally confirmed by the Japanese attack on Pearl Harbor on December 7, 1941. The Japanese had been powering ahead: from the late 1920s, submarine development had become an intrinsic element of their growing front-line naval capability.

While the Navy struggled to recover from the Pearl Harbor attack, the submarine force took the war to the enemy. Operating from Pearl Harbor, and Australian bases at Fremantle and Brisbane, and eventually employing the reliable new Gato, Balao and Tench classes, the submariners began to fight back in some style. By the late summer of 1942, the scorecard was beginning to show signs of improvement but the Japanese still held the upper hand by far, and were inflicting heavy losses on the US Navy. In August, the aircraft carrier *Saratoga* was torpedoed and knocked out of the war for months. Three weeks later, the Japanese submarine *I-19* sank the aircraft carrier *Wasp* and seriously damaged *North Carolina*, America's newest and most powerful battleship. *Wasp*'s sister carrier *Hornet* was hit in October, a crippling blow at the time when the Battle of Guadalcanal hung in the balance.

ABOVE: **USS *Plunger*, sister submarine of *Pompano*, survived the war and members of the crew display her battle-flag, while the sailor seated in the centre is wearing a Japanese sailor's hat. The photograph was dated June 21, 1943, following *Plunger's* sixth war patrol.**

Apart from the hasty building programme, the Americans also had to make considerable readjustments to methods of attack, given that most training and operational work had been planned around the principal role of coastal defence. Across America, too, scientists and mathematicians were working around the clock to produce new technology for submarine detection.

One of the leaders in new methods of tactical warfare that became common practice emerged among submarines operating from the Brisbane base, under the command of Admiral James Fife. He suggested a trial of a new command and control system in USS *Wahoo* under Lt Commander Dudley W. Morton. Executive Officer, Lt Richard O'Kane, manned the periscope leaving Morton free to evaluate the entire combat situation. On her next patrol, *Wahoo* sank 32,402 tonnes/31,890 tons of Japanese shipping. Morton received the first of four Navy Crosses and his ship took home a Presidential Unit Citation. Later in the war, as commanding officer of USS *Tang*, O'Kane received the Congressional Medal of Honour and became the Submarine Force's leading ace of the war, credited with destroying 31 ships totalling 227,800 tonnes/224,202 tons.

Similar initiatives were introduced at the Fremantle base under Admiral Charles Lockwood. He totally revamped tactics employed by the newer submarines and towards the year's end, America's submariners slowly began to turn the tide, and ended 1942 with 180 Japanese ships sunk, totalling 725,000 tonnes/713,550 tons. In fact, it was a meagre tonnage compared to the U-boat tally for the same year – 1,160 Allied ships of more than 6 million tonnes/5,905,239 tons sunk.

The impetus continued apace as the new American submarines piled into the region, and after the recapture of Guam in August 1944, boats based there and at Saipan succeeded in imposing a virtual blockade against Japan, causing massive shortages in oil, raw materials and food. By the autumn of 1944, Japan's remaining five fleet carriers had been sunk, and the sea war in the region was as good as settled.

TOP: **A typical scene in the control area of a US submarine before the action gets hot, but the apparent calmness belies the fact that at any moment chaos and confusion may ensue. This was especially so when the US submarine chiefs raised the stakes early in 1943 by totally overhauling tactics, and continued to do so as the war progressed. They employed more aggressive daytime submerged and night-time surface attacks, partly enabled by improving technology, which resulted in considerable damage to the Japanese fleet.** ABOVE: **A periscope view of the carnage on the surface after a successful attack upon enemy shipping, indicating whether it was safe to surface.** LEFT: **The crews of all American boats kept track of progress with their battle flags, this one from USS *Besugo*, a *Balao*-class submarine, circa early 1944.**

# Clandestine travellers aboard

Submarines in all theatres of World War II were used by all sides for many special operations. The Germans landed agents in Britain, America and elsewhere. The British did the same across Europe and the Far East, and to the last, submarines were in continuous use for clandestine operations. Indeed, some of the most famous exploits of particular heroism, which resulted in many a post-war movie, were launched from submarines, including sabotage operations by the Special Boat Service, assault operations by Commando units, the famed but ill-fated Cockleshell heroes and members of the Special Operation Executive. Submariners were also the carriers of a top secret organization whose very existence was denied for 15 years after the war because of ongoing operations on sensitive coastlines – that of the Combined Operations Pilotage Parties (COPPS). This was formed in great secrecy in September 1942 for harbour and beach reconnaissance prior to landings of Allied forces in virtually every theatre of war, operations which still exist today in the shape of Special Forces the world over.

More than 40 Royal Navy submarines were involved in clandestine operations in World War II, and several of them, including the T boats in the Mediterranean and the brand new S-class boat of Lt Norman "Bill" Jewell, commissioned in 1942, were used repeatedly for the service. The British submarine service was a forerunner of these operations and gained early practice in 1940, carrying Commandos to the German-held Norwegian coastline, bristling with all kinds of anti-submarine vessels, mines and air attacks. Lt Commander A.C. Hutchinson and the crew of *Truant* experienced the fiercest and most sustained attacks by depth charges ever known by a British submarine up to that point of the war while carrying his Commando passengers.

Next, the early exploits of the fledgling SBS began in the Mediterranean, when teams were carried around by submarines to be landed clandestinely on Italian shores to blow up railway lines, viaducts, water supplies and bridges. The most famous duo of Lt Robert "Tug" Wilson, an architect from Leamington Spa, and his partner, Marine Wally Hughes, led the way, hitching rides in *Urge* with Lt Commander Tommy Tomkinson. They would be floated off the casing of the boat in their canoe, perhaps 4.8km/3 miles offshore. They would then paddle their little craft ashore, loaded with explosives, where the two men would carry their load to the designated target, set the charges and make a hasty exit. They would lay up until the appointed time, when their host submarine would (hopefully) reappear to pick them up, mission completed.

There were many, many dangerous moments when these operations were interrupted by German and Italian aircraft or surface ships. In one ill-fated operation in November 1941, two T boats, *Talisman* and *Torbay*, were put at great risk to insert 56 Commandos on to a beach 322km/200 miles inside enemy lines on a most dangerous mission. Shortly before British forces launched an all-out attack to gain control of the North African coast from German and Italian forces, they were assigned to assassinate Field Marshal Erwin Rommel, then

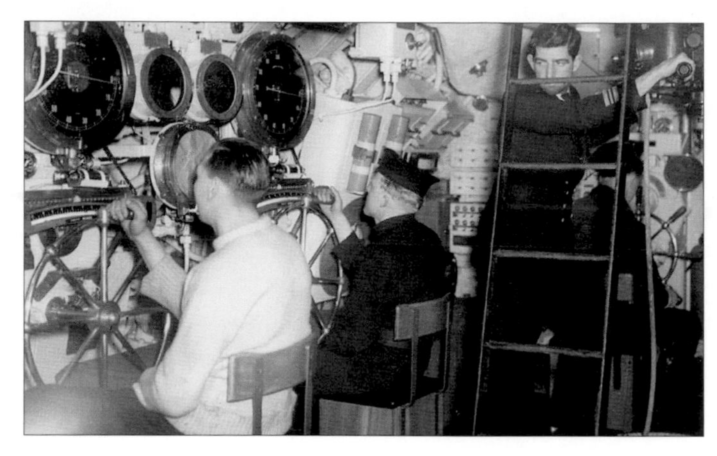

**ABOVE: The control room of HMS *Tuna*, with Lt Norman at the periscope of the boat that launched "The Cockleshell Heroes", 12 Royal Marines who made their famed canoe-borne attempt to blow up German ships in the Gironde estuary. LEFT: The Anzio landings, part of the Allied invasion of Italy. The beaches had been presurveyed by members of the top-secret Royal Navy Combined Operations Pilotage Parties in their canoes, and clandestinely delivered by submarine.**

FAR LEFT: **The T-class submarine *Triumph* that delivered many of the secret operatives of organizations such as the Special Boat Squadron or the Special Operations Executive to their destinations on dangerous missions behind enemy lines.** LEFT: **Folboats or canoes used by the SBS and Navy frogmen were stowed aboard the submarines and unloaded near the target coastline.** BELOW: **The canoeists would then paddle their way into shore, and then, in most cases, the submarine would return at a pre-determined time to collect the saboteurs.**

ABOVE: **The American Navy paid particular interest to early British successes in clandestine operations and set up their own units. Although hardly invisible, the USS *Nautilus* was used on early missions, and sailing with submarine *Argonaut*, carried Marines of the 2nd Raider Battalion to stage diversionary attacks on Japanese-held territory.**

commander in chief of Axis forces, who was supposed to be at his inland villa. In very rough seas, *Torbay* and *Talisman* were heaving as the inflatable boats were launched towards the shore. It was so choppy that only 45 made it off the submarines. It was a heroic but ultimately unsuccessful mission. Rommel, it turned out, was not at the villa but in Rome. The Commando force met heavy resistance, and the submarine commanders risked their own safety in waiting around at the rendezvous to pick up the men who, in fact, never returned. Only four Commandos survived the mission and only evaded capture by walking back to their base across a hostile landscape, a journey that took 40 days.

One of them was Bombadier Brittlebank of the Royal Artillery, who subsequently replaced Marine Hughes, who was ill, as Tug Wilson's partner for further missions off the Italian coastline. Their exploits were, however, to be curtailed. The danger of these operations to the submarines was demonstrated by the fate of *Triumph*, under Lt Commander W.J.W. Woods. She carried out a number of special operations, landing commando teams all around the Italian coastline until she was assigned to land a unit near Athens, sailing from Alexandria on December 26, 1941. Four days later Lt Commander Woods sent a signal confirming the landing of the party, and *Triumph* was due to return to pick up the commandos two weeks later, but failed to make the rendezvous. Nothing further was heard of the submarine.

Another of the famous submarine-launched sabotage groups were the canoeists of Operation Frankton, more commonly known as The Cockleshell Heroes, the brainchild of Lt Colonel H.G. "Blondie" Hasler. Pairs of men were carried by the submarine *Tuna* to the mouth of the river Gironde, on the west coast of France. Five pairs then paddled 113km/70 miles upstream by night to attack 12 German ships lying in the supposed safe haven in the Bassens-Bordeaux area. However, only one pair, Hasler and his colleague Ned Sparks, survived the mission. Of the remaining eight men involved, six were captured and shot as saboteurs by the Germans and the remaining two were lost, presumed drowned.

# "The most important capture of the war"

TOP: **HMS *Bulldog*, the most famous of the seven Royal Navy ships to bear the name, was a Class B destroyer launched in 1930, and she remained in service until 1946 when she was scrapped. The heroic actions of her crew in capturing the first Enigma machine undoubtedly helped change the course of the war.**
ABOVE: **The submarine from which the Enigma machine was extracted, *U110* was captained by Fritz-Julius Lemp, a commander of some notoriety, notably for sinking the liner *Athenia*.**

Kapitänleutnant Fritz-Julius Lemp, the U-boat captain who sank the *Athenia* on the first day of the war, had been a busy man, commanding *U30*, the lead boat in the 2nd U-boat Flotilla. By May 1941, he was a holder of the Knight's Cross and commander of *U110* in the wolf-packs scouring the Atlantic for their prey, operating in partnership with *U201*. On May 9, Lemp sank two ships totalling 10,000 tonnes/ 9,842 tons, but when he came up to confirm his final "kill", he left his periscope aloft too long and was spotted by the convoy escort, the corvette HMS *Aubretia*. The escort gave the alert to other ships while she herself unloaded a massive bombardment of depth charges.

*Aubretia* was joined by the destroyers *Bulldog* and *Broadway*, and their attack was delivered with such force that Lemp was forced to surface. As he came up, a dozen men on *U110* rushed to man the guns but were themselves shot by the waiting British ships. Lemp also discovered that *Bulldog* was already lined up for a ramming operation, which her commander Commander Joe Baker-Creswell had no intention of carrying out, having already considered the possibility of a capture. Lemp, expecting the ramming to arrive any time soon, ordered his men to abandon ship, apparently assuming that once his boat was sunk, her secrets would go to the bottom with her. This boat did, indeed, contain secrets – none other than one of the famous Enigma cipher machines, as well as a vital codebook on how to use it.

So, over the side went the surviving members of the crew, Lemp with them, but he never made it to the British ships which picked up the survivors. Many said after the war that he was shot in the water by a British officer who led the boarding party from *Bulldog* to take possession of *U110*. David Balme, the man accused of this act by a German author after the war, denies this categorically in his memoir held in the Imperial War Museum's Sound Archive, and said Lemp died in some other way. The speculation was that realizing his boat, a IXB class, was not going to sink, he either tried to swim back to save the secrets aboard and drowned, or committed suicide, realizing his cardinal sin of failing to scuttle his boat.

*Bulldog*'s crew, meanwhile, had pulled alongside the U-boat and began stripping her of everything they could carry. David Balme recalled: "I was in the control room when our telegraphist called me over and said '*Look* at this, sir,' pointing to what appeared to be an old typewriter. We both thought it looked interesting, and that we'd better take it over. And that is how we discovered the Enigma cipher on board *U110*."

ABOVE LEFT: **Another Enigma-carrying boat, *U505* was captured by the Americans exactly three years after *U110* was taken by the crew of HMS *Bulldog*, but because the British government had ordered a clampdown, which remained in force for 30 years, it was widely believed that *U505* was the first to be captured – a belief wrongly portrayed in a Hollywood movie.** ABOVE: **U-boat pens provided a safe haven for the German submarine fleet around the French coast as well as in Germany. These were especially vital at the time of the Battle for the Atlantic.** LEFT: **A typical interior of the boats operating during that era.** BELOW: **The Enigma machine, whose mysteries were unlocked at Bletchley Park, thus providing the British with an incredible insight into forthcoming Nazi operations.**

Along with it were masses of charts, maps and codebooks – the whole shooting match that revealed the Enigma code, as well as charts of German minefields. David Balme also discovered what became known in the Royal Navy as the grid map – a remarkable map of the Atlantic, divided into squares with lists of U-boats assigned to certain areas.

It took hours to ferry the material back to *Bulldog* and then Baker-Creswell took the boat under tow. However, as soon as the Admiralty received a description of what had been found, they ordered the boat to be scuttled at once, and all the prisoners taken to Iceland, where they would be interned. The whole operation was now very cloak and dagger to ensure that no one – especially not the German High Command – should know the U-boat and her secrets were in British hands.

An expert from Bletchley Park met *Bulldog* when she arrived in Scapa Flow and took the Enigma machine and the books away. Every man on board the ship was sworn to secrecy. The find was enormously significant and helped the Ultra code-breakers to read top secret communications between Hitler and the German High Command for months ahead, a fact that Churchill did not reveal even to Roosevelt until the end of 1942.

The capture of *U110* was one to become one of the best kept secrets of the war, and indeed was not publicly revealed until 30 years later. That is why the American Navy put into printed history the claim that it captured the first U-boat containing dramatic secrets. How were they to know the British had already done it! Their capture came in May 1944 when US Navy Task Group 22.3 sailed from Norfolk, Virginia, for an anti-submarine patrol near the Canary Islands. On June 4, USS *Chatelain* reported a sonar contact with a submerged submarine. Two fighter planes from the USS *Guadalcanal* were called down to fire their guns into the water to help mark the location of the submerged *U505*. The USS *Chatelain* then fired a pattern of 14 depth charges, forcing *U505* to surface. While the USS *Chatelain* and the USS *Jenks* picked up survivors, the USS *Pillsbury* sent its whaleboat to the *U505*, where Lt Albert L. David led a nine-man boarding party to capture the submarine.

Page 2. Issued in lieu of No 09650 lost. Page 3. Navy Form S.1511

NAVAL IDENTITY CARD No. 148228

Surname MARTIN

Other Names WILLIAM

Rank (at time of issue) CAPTAIN, R.M. (ACTING MAJOR)

Ship (at time of issue) H Q COMBINED OPERATIONS

Place of Birth CARDIFF

Year of Birth 1907

Issued by

At ADMIRALTY

Date 2nd February 1943.

Signature of Bearer W. Martin

Visible distinguishing marks NIL.

LEFT: **The fake identity documents secreted in the pockets of "Major Martin", which helped persuade German intelligence into accepting that he carried genuine details of plans for an Allied invasion of the southern coastline of Greece when, in fact, they were to descend upon Sicily.** BELOW: **US Paratroops join the Allied invasion of Sicily, a task greatly facilitated by the fake documents found on the body launched from HMS *Seraph*.**

# Operation Mincemeat (or "The Man Who Never Was")

Lt Commander Norman Jewell had already carried out numerous special operations along the coast of North Africa since taking delivery of the new S-class submarine *Seraph,* including the covert delivery of US General Mark Clark to Algeria in October 1942 to try to persuade the Vichy French to support the planned North African landings. A month later, Jewell and his crew provided the transport for the rescue of French General Henri Giraud off the south coast of France after he had escaped from a German prison camp. It was a dangerous episode because the general's arrival was delayed for almost a week, but *Seraph* had to remain on station, diving and surfacing in hostile waters, ready to pick up her passenger.

The most famous of *Seraph's* secret exploits subsequently became popularly known as "The Man Who Never Was": a cunning scheme codenamed Operation Mincement, aimed at fooling the Axis powers into thinking that the Allied landings in Sicily would, in fact, take place elsewhere. Norman Jewell tells the story personally, in his memoir for the Imperial War Museum sound archive:

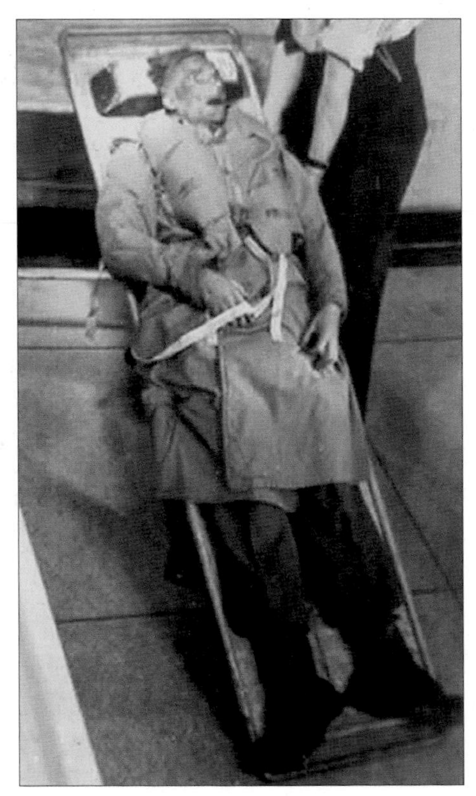

*"I was told to report to naval intelligence and there met up with a team concocting an incredible scheme…A body was to be washed up on the coast of Spain which would carry in an attaché case secret papers, a letter for Eisenhower, the Commander-in-Chief in North Africa, saying that rather than going for Sicily the invasion would be on the south coast of Greece.*

*"My knowledge was limited exactly to that which I needed to know in order to carry out my part of the plan. However, we spent a good deal of time building up a background for this body, who was to be called Major Martin, providing him with a life, a girlfriend, and in his pockets theatre tickets, restaurant receipts, a letter and so on. They knew it would be checked out by German spies in London. Then (when the time came), the body of Major Martin was placed in a canister about the size of a torpedo.*

*"The body was packed in ice so that it would be the right age when washed up. We were given clear passage down to*

LEFT: **The fictitious Major William Martin – "The Man Who Never Was" – being cast into the sea off the coast of Spain.**

LEFT: **HMS *Seraph* was one of the improved British S-class boats, commissioned in 1942.** BELOW LEFT: **One of the restaurant bills found in the pockets of the dead man that also helped convince the Germans as to his authenticity.** BELOW: **On their way to the true target, the massive task force of British and American troops join the invasion of Sicily by sea and air.**

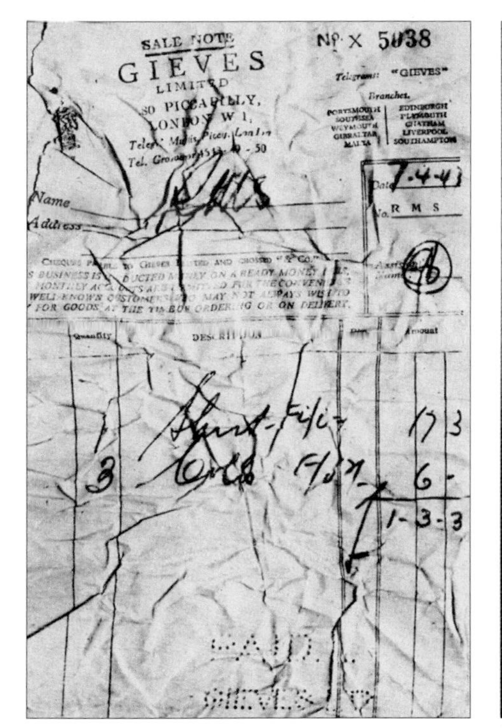

SALE NOTE No X 5038
GIEVES
LIMITED
80 PICCADILLY,
LONDON W 1.
Telegrams "GIEVES"
Branches.
Name
Address
Date 7-4-43
R M S
Quantity DESCRIPTION Amount
1 17 3
3 6 -
1-3-3
PAID
GIEVES

the coast of Spain, which meant that the RAF was aware of our presence. We were also told that we would not be required to attack anything en route because our mission was too important. However, we were taken out of our safe passage line to attack some shipping going to Bordeaux. We never saw them, but thanks to that instruction we were then bombed three times by the RAF. We eventually arrived off the coast of Spain at the point where Spain joins Portugal. Until that point, no one on board – apart from myself – knew that the canister contained a body.

"The cover story was that it was carrying meteorological equipment to fill in the reporting gap for the Meteorological Office in London. But now, the body was to be retrieved from the canister to be put over the side. I had to tell the officers because I could not do the whole thing on my own. I did not give details of the plan, but in any event I had to swear them to secrecy. They were surprised; some were shocked...dispatching the body of an unknown man into the sea in this manner. We took off the end of the canister and brought the body out. It was the first time I had seen it. We made a final check to make sure that his papers and attaché case containing the secret 'invasion' plans were intact and attached to his wrist, and then slid him over the side. We went full astern on the motor so that he would be pushed on his way. We said a few words as a funeral service over him and then secured the canister on the casing and dived. At daybreak we surfaced to try to get rid of the canister and we had a hell of a time. It had been so designed to keep the ice from melting that it had air pockets all the way around it, and even though we had put about 200 bullet holes through it from a machine-gun nothing happened at all. We then had to go alongside it again and put some plastic explosives inside and outside and then withdrew while it blew up. It then disappeared finally. We then went on to Gibraltar, arriving the next evening. As we arrived, someone came over the gangway carrying a telegram for me, which said the parcel had been received: so the body had arrived safely.

"As would be discovered later, the Germans went to great lengths to discover the background of Major Martin through their spies in London and could not discredit the information that we had placed about him. In due course, the Germans withdrew an armoured division and sent it through Italy and down the other side towards Greece. They were at least a division less in Sicily when the landings eventually took place."

In fact, the ploy was so successful that the Germans thought Sardinia and Greece were the intended objectives until well after the landings in Sicily had begun. As for the "body" that became Major Martin, it was reported in 1997 that it was that of a homeless alcoholic from Wales named Glyndwr Michael, although even that claim has since been challenged. Whoever it was, the body was buried in Heluva, Spain, with a gravestone in the name of Martin.

LEFT: **The Italians were masters of the human torpedoes and Chariots and had produced a working prototype, which was tested at La Spezia in 1936. This ultimately led to the construction of submarine-borne containers to deliver them to the target area.** BELOW: **The British quickly caught on and introduced their own charioteers, here in rehearsal for the invasion of Sicily.**

# The chariots of stealth

The Japanese, Italians and latterly the British all had specialist teams working on the development of underwater guerrilla warfare. The Axis navies were well ahead of the field when war came as a result of experimental work in the 1930s. After the exploits of the SBS and the Cockleshell Heroes, Lord Mountbatten, as head of Combined Operations, sent a memo to the Chiefs of Staff, pointing out that the men's determination and courage demonstrated a good example of the successful use of limpeteers.

He already had a report which indicated that Italian underwater activity was "filled with ideas involving gadgetry and motorisation and experimentation". In fact, the Italians were well ahead of everyone in their perfection of a number of devices and submarine attack apparatus that would become the model for the future. These included effective breathing apparatus for underwater swimmers, double limpets, human torpedoes, piloted torpedoes, miniature torpedoes and exploding attack boats. Teams of dedicated and highly trained men were inspired by the famous Commander Belloni and the infamous Prince Julio Valerie Borghese from the 10th Light Flotilla of the Italian Navy.

This particular skill of the Italians really began to worry the Allies in 1941, at a time when the British fleet was reduced to two battleships in the Mediterranean, HMS *Valiant* and the brand new HMS *Queen Elizabeth*, which lay sheltered behind torpedo nets at Alexandria. At 03:30 hours on December 19, two Italians were discovered clinging to the anchor buoy of *Valiant*. They surrendered immediately and were taken ashore for interrogation and then, to their dismay, back to *Valiant*, where they confessed that the battleship was about to blow up. The crew was mustered on deck and the watertight doors were closed, but shortly after 06:00 hours the ship rocked and shuddered as the charge set by the two Italians blasted a large hole in her stern. Soon afterwards, *Queen Elizabeth* reared up from two explosions from charges attached below the water line, and both ships were temporarily out of the war.

It soon emerged that they had been attacked by three human torpedoes known as Maiali (sea pigs) driven by a team of six men from the 10th Light Flotilla, trained to remain under water for miles wearing flexible rubber suits, breathing gear and fins. They had been launched from Prince Borghese's submarine *Sciré* off Alexandria, two astride each of the human torpedoes. The operation even drew praise from Churchill as an example of "extraordinary courage and ingenuity".

Mountbatten drew together a number of eminent military and scientific experts to look at new weapons and methods for his men to "study, coordinate and develop all forms of stealthy seaborne attack by small parties and pay particular attention to attacking ships in harbour". Among the projects launched immediately was the Chariot, a hefty torpedo-shaped submersible which was a copy of the two-man Maiali used by the Italians (one had been captured off Crete). Its crew, equipped with breathing sets that would allow six hours of diving, sat in the open astride the Chariot, which could travel at about 16 knots and carried one 272kg/600lb charge. They were hazardous and slow. The first torpedo force came into operation in September 1942, under the auspices of the submarine service, in a plan to attack *Tirpitz*, the largest and strongest warship ever built in Germany, at 53,000 tonnes/ 52,163 tons.

She was even more powerful than her elder sister, *Bismarck*, and was based in occupied Norway from January 1942. From there, she and her supporting ships would sally forth to attack Allied shipping.

The British began to consider every scheme to destroy the German monster, and in October 1942, an audacious plan was launched to sink her using the Chariots. Two of the Chariots were hidden in a double bulkhead of a Norwegian fishing trawler which would then proceed to Trondheim; there, they would be released with crews to attack *Tirpitz* in her berth behind anti-submarine nets in a fjord north of Trondheim. Although the trawler managed to get through three German checkpoints, the expedition was hit by a sudden change in the weather as the Chariots were being launched. They had to be abandoned, and the crews swam ashore and escaped into Sweden. No further attempts could be launched because of the low temperatures that year.

More than 50 Chariots were built, and their crews experienced mixed fortunes. Their best results were in the Mediterranean, although they were never really considered a success. Operation Principal, for example, was an attack on Palermo harbour on January 3, 1943. Five Chariots were used. Each, with a crew of two, was embarked on the deck in watertight containers in the submarines *Trooper* and *Thunderbolt* (ex-*Thetis*). *Unruffled* took part as the recovery boat after the operation was complete. One Chariot sank the Italian cruiser *Ulpio Traiano*, and another damaged the liner *Viminale*, which sank later in the year. Of the remaining three, one broke down en route and was picked up by *Unruffled* six hours later, a second sank through unknown causes, and on the third, the driver ripped his diving suit negotiating harbour defences and drowned.

ABOVE LEFT: **Luigi de la Penne, one of the pioneers of Italian charioteers, was one of a team of six that attacked British ships in Alexandria harbour in December 1941, causing damage to two British battleships, *Queen Elizabeth* and *Valiant*. He was captured but he refused to inform the ship's captain of the mines attached to the ships' hull until a few minutes before detonation, thus allowing the British to evacuate. Although imprisoned, he worked for the British as a frogman after the Italian surrender.** ABOVE RIGHT: **The British version of the charioteers, whose operations came to the fore at the time of the Allied invasion of Sicily and elsewhere in the Mediterranean.** LEFT: **A British Chariot being unloaded for operations.** BELOW: **The altogether more successful Italian version of charioteering.**

# Giant tasks for midget submarines

While Chariots were something of a haphazard affair, midget submarines were a different story altogether. Various nations had experimented with the small boats that could be transported and utilized in situations that their giant sisters would have little chance of even approaching, let alone manoeuvring into position for an operation. The Japanese and the Italians were well ahead of the field in design and numbers, the former producing several dozen in different classes and sizes. By the beginning of the war, the Japanese were already well advanced in their use, also with several dozen in operation, beginning with the A class, 23.9m/78ft 6in long and displacing 46.7 tonnes/46 tons submerged, and capable of a tidy 23 knots on the surface and an incredible 19 knots submerged.

They were originally intended to be carried by larger submarines or surface vessels and deployed to attack an enemy fleet, but they discovered that the midgets were particularly useful for special operations against ships in enemy harbours, where they could be transported and released in packs. Five of the A-class boats took part in the attack on Pearl Harbor and were evident again in May 1942 in raids on Sydney, Australia, and Diego Suarez in the Indian Ocean. Notably, however, they were well used against American ships at Guadalcanal in 1942–43. Japanese midgets in the A, B and C classes were in continual production, ending with the Type D boat, *HA-77*, completed in January 1945. Some 115 units had been completed when Japan capitulated in August 1945, and nearly 500 more were under construction.

Japan also produced some special attack midget submarines which were actually manned torpedoes. The highly successful Type 93 Long Lance torpedo was lengthened and adapted by inserting a central cockpit for the pilot. They became the Kaiten midget submarine that could reach speeds of up to 35.4kph/22mph, and up to six could be piggy-backed on conventional submarines for delivery to the attack area.

Meanwhile, the Italians were in production with a stylish-looking craft in their CB class as the war began, and it was this model that provided additional tips for both the British and the Germans, who took possession of a couple of their boats operating in the Mediterranean. In fact, the British, the

ABOVE: **The Japanese Navy was first in action with midget submarines, launching five Type A models in the Pearl Harbor raid of December 7, 1941, transporting them piggy-back on larger submarines and launching the night before the attack, and thereafter throughout the war.** LEFT: **The British X-craft, which had a varied, terrifying but VC-winning history of high-profile operations, including an attack on the German battleship *Tirpitz*.**

LEFT: **In mid-1944, the Japanese Navy developed the Koryu Type D midget coastal defence submarine for a five-man crew, two more than in the Type C.** ABOVE: **Although classed as a human torpedo, this version of a Japanese *Kaiton* was more of a mini-submarine, with a large conning station attached.**

Americans and the Germans both came late to this field of development, the latter eventually launching a massive programme of midget development, building six different classes, ranging from one-man torpedoes to seven-man midgets. In all, the Germans had planned to produce over 1,700 of such craft, the first appearing early in 1944, but the war ended before many of that number were actually commissioned.

The Royal Navy's own prototype X-craft was laid down in 1939, when intelligence reports gave inspiring accounts of foreign developments, but the British and the Americans had already experienced attacks from midget craft at Pearl Harbor and in the Mediterranean before the British version was commissioned in October 1942. It was classed as *X3* (the first two Xs were already taken: *X1* was an experimental boat built in 1925, and *X2* was a captured Italian submarine which had given the British a few extra pointers). A second prototype vessel, *X4*, was constructed and based on the first British boat and the Italian version. Thereafter, rapid production began with six X-craft (*X5* to *X10*) being built by Vickers, initially for deployment in European waters. The British midgets made their name in some spectacular special missions and, later in the war, when they were used for risky surveillance and mapping of landing sites on hostile coastlines.

Their most famous mission was the attack on the German battleship *Tirpitz*, with which Churchill had an obsession. This magnificent and powerful vessel represented a major threat to British shipping and he wanted it out of the war. The Germans, well aware of British intentions, kept the ship Altenfjord, Norway, out of harm's way, which was also the berth of the Germans' other heavy weapons, the battleship *Scharnhorst* and the pocket battleship *Lützow*. They were the targets in September 1943 when six X5-class submarines, each crewed by four highly trained operatives, set out on a mission that is today a classic point in submarine warfare.

The X-craft, unlike the high-speed Japanese models, had electric motors that generated just 30hp and a 12-knot top speed when submerged. So they had to be towed towards their quarry by regular-sized submarines. Even so, the outward journey to Norway of 1,770km/1,100 miles took eight days in often heavy seas. En route, *X9* (which had been detailed to attack *Scharnhorst*) sank and *X8* (heading for *Lützow*) had to be scuttled, leaving the four remaining X-craft to cover the last 80.5km/50 miles under their own power through the treacherous fjords, mine fields and anti-submarine nets that were in place on the approach to the *Tirpitz*.

ABOVE: **The German *Neger* consisted of two linked G7e torpedoes, the upper with the warhead removed and a transparent driver cockpit installed; the bottom torpedo would be released.** BELOW: **The 11.2 tonne/11 ton one-man German Molch midgets were all electric, designed for submerged coastal operations, with a range of 64.4km/40 miles, carrying two torpedoes. Around 390 were built.**

On September 22, 1943, *X6* (under the command of Lt Donald Cameron) and *X7* (under Lt Basil Place) followed an old freighter through this obstacle course and reached their target. *X6* was spotted, and was under heavy fire from the surface. When Cameron attempted to make his escape, he became caught in the nets but still did not give up. As he surfaced, he dropped two explosive charges and then managed to scuttle the boat before leading his crew to surrender. *X7* also released two charges beneath *Tirpitz* but ironically was damaged when the explosives left by *X6* detonated, and Basil Place also had to surrender. Two of his crew did not survive.

Meanwhile, *X10* under the command of Lt K.R. Hudspeth also managed to sneak into the inner fjord, only to be hit by mechanical trouble. Rather than give up and possibly alert the Germans to the presence of a larger force, he remained hidden for five days before he could return to the towing ship. The boat had to be scuttled on the return journey to England. A third submarine, probably *X5*, was sighted by Cameron close to its target but came under heavy fire and was never seen again.

ABOVE: **The aft portion of the Kaiten (Return to Heaven) Type 1, of which 300 were produced. It became the first Japanese Special Attack weapon whose operational use involved the certain death of the crew and preceded the Kamikaze aircraft.**
BELOW LEFT: **The Biber (Beaver), the smallest German submarine built to attack coastal shipping ahead of the Allied invasion of Europe.**

In spite of this set-back, the raid on *Tirpitz* herself was successful, and that was the key object of the exercise. Although the explosive charges left by the midget did not have the power to sink her, they did sufficient damage for the vessel to be put out of the war temporarily, and had to be towed south for repairs. Apart from an actual sinking, this could not have been a better result because *Tirpitz* was now in the range of RAF and Fleet Air Arm bombers and over the coming months, after several attempts, she was finally blown apart when her own ammunition store was hit by British bombers. Fewer than 90 of her 1,000 crew survived. The role played by the midget submarines in crippling *Tirpitz* was recognized in 1944, when Basil Place and Donald Cameron were awarded the Victoria Cross.

Meanwhile, new versions of the British midgets were coming off the production line, beginning with the XE class which carried out a number of successful raids around Europe, including that of *XE3* under the command of Lt Ian Fraser, which put a large floating dock at Bergen out of action for the rest of the war. As targets began to become fewer, six of the *XE*-class boats were shipped for operations in the Far East for what turned out to be the final months of the war. They were earmarked to carry out raids against the Japanese at a number of locations, including Singapore and Hong Kong. Again, a

LEFT: **An aerial view of the German battleship *Tirpitz*, which was damaged in an attack by the British X-craft midget submarines. This allowed British bombers to sink her.** ABOVE: **One of the most successful German miniature submarines was the Seehund Type 127, of which the Kriegsmarine had planned to build over 1,000 from 1944. They had a crew of two, carried two under-slung G7e torpedoes and had a range of 300km/186 miles at 7 knots.**

number of successes were chalked up, including that of *XE4* under the command of Lt M.H. Shean, of the Royal Australian Navy Volunteer Reserve, whose team managed to cut the Hong Kong to Saigon cable and the Saigon to Singapore cable.

The most famous of the missions, however, was reserved for Ian Fraser's *XE3*. His mission was a *Tirpitz*-style raid on the Japanese heavy cruiser *Takao*, moored in the Johore Strait, Singapore. Once again, the X-craft was towed towards its target location, and then continued on under her own steam. They reached the *Takao* without incident and settled beneath her to get to work. Leading seaman James Magennis, the diver, went out and attached six limpet mines to the hull of the ship, a task which took forty minutes because he had to scrape each position free of mud and barnacles. He returned aboard exhausted. A further problem occurred as Fraser attempted to turn away and escape before the explosion. To his horror, the boat would not budge because, while working on the explosives, the tide had gone down and the boat had become stuck in a very narrow hole. Only by going ahead and then full astern for several minutes did he manage to escape, only to encounter a further problem: an empty limpet container had jammed, preventing the side cargoes of explosives from being jettisoned, which gave the submarine an uncontrollable list. Magennis volunteered to go back outside, and spent 15 minutes freeing the load. *XE3* eventually made her escape and, while returning to base unscathed, the charges detonated to put the 11,177-tonne/11,000-ton cruiser on the bottom. Lt Fraser and Leading Seaman Magennis were both awarded the Victoria Cross for their bravery in performing this operation.

Although disliked by some, the X-craft had performed a number of exceedingly useful tasks, including their use by the top-secret group Combined Operations Pilotage Parties (Copps) in preliminary surveys ahead of troop landings and, notably, to mark the invasion beaches for the D-Day invasions.

ABOVE: **The Kairyu (Sea Dragon) was another Kamikaze midget submarine for a two-man crew and was fitted with an internal warhead for suicide missions. Designed at the beginning of 1945 to meet the anticipated approach of invading American naval forces, 200 had been built by the time of Japan's surrender.** RIGHT: **Various methods were tried to disguise the tell-tale periscope of midget boats; this was the Beaver camouflage.**

LEFT: **The forlorn sight of U-boats in Hamburg harbour as Germany capitulated, a reflection of the final days of the once-mighty Kriegsmarine.** BELOW: **The experimental *U3008*, an XXI submarine, known as the electro-boats, which provided technology for the Allies to plunder. Transferred to Portsmouth, New Hampshire, she was commissioned into the US Navy.** BOTTOM: **Having the distinction of being the first German submarine to surrender at the war's end, the crew of *U249* come ashore at Weymouth having shown the white flag to HMS *Magpie* off the Cornish coast.**

# The cost and the legacy

For the second time in the 20th century, the final reckoning of submarine warfare produced a devastating catalogue of destruction of hardware and human life, as well as a record of many thousands of incidents of great heroism and skill, regardless of the flag under which the men were fighting. The U-boat force was undoubtedly the greatest in the world in terms of quality and innovation. It also suffered the highest proportion of casualties. A total of 1,154 diverse and versatile U-boats were commissioned prior to or during World War II, not counting the 135 still under construction at the end of the war and excluding the midget submarines. However, the toll had been relentless: 727 U-boats manned by crews totalling 26,918 were lost, the majority sunk by Allied surface ships and shore-based aircraft. The progression of those U-boat losses paints a mini-portrait of the way the fortune of war proceeded: 1939: 9 lost with 204 crew. 1940: 24 lost; 643 crew. 1941: 34 lost; 887 crew. 1942: 85 lost; 3,277 crew. 1943: 235 lost; 10,081 crew. 1944: 219 lost; 8,020 crew. 1945: 121 lost; 3,806 crew. The human cost was enormous, and the final figures showed that Germany lost an astonishing two-thirds of the men who served in the submarine force.

The British, with far fewer submarines, lost 74 boats during the war years, manned by crews totalling 341 officers and 2,801 ratings. In addition, 50 officers and 309 ratings became prisoners of war. America lost 52 submarines and of the 16,000 men serving in the US submarine force, 375 officers and 3,332 men did not survive. For all her productive pre-war promise, Japan's submarine service did not perform as well as her enemies'. Excluding midgets, Japan started the war with 63 ocean-going submarines and completed 111 during the war, giving a total of 174, of which 128 were lost during the conflict. Japan's emphasis on size proved to be a disadvantage, in that the huge boats were easy to sight and target. They were slow to dive, hard to manoeuvre underwater, easy to track, and easy to hit, especially by the US Air Force. Japanese submarines also had no radar until the first sets were installed in June 1944. Russia had 272 submarines in operation during the war and lost 108. Details of their operations are scant because Soviet submarines were rarely on patrol. However, it is known that one of their boats caused the greatest disaster in maritime history, in sinking the German liner *Wilhelm Gustloff* on March 30, 1945. The liner was moving more than 6,000 German troops and up to 2,000 civilians from Poland ahead of advancing Russian forces. She was sunk soon after leaving the port of Gotenhafen (Gdynia), and fewer than 1,000 people were rescued.

LEFT: **Surrendered VIIC boats, the mainstay of the German force from 1941 and still in production at the end of the war. They bore the brunt of the Allied anti-submarine onslaught from late 1943 onwards. One of them, *U96*, was featured in the film *Das Boot*.**

BELOW: **The revolutionary snorkel fitted to the XXI types that could have won the war, had they been in production earlier.**

At the end of it all, as in World War I, the victors picked over the spoils. There had been a resurgence of successful U-boat activity towards the autumn of 1944, when freshly launched German submarines were fitted with an important new invention: the snorkel ventilating tube. First tested by the Dutch Navy and picked up by German scientists, the retractable device contained air-intake and exhaust pipes for the engines and general ventilation. The telescopic tube fed air into the diesel engine and carried off exhaust gases. A U-boat equipped with the snorkel could run submerged on her diesel engine while also charging her batteries; while the boat could do so only at a slow speed and at shallow depth, the chances of being picked up by radar were greatly reduced. When they heard of it, Allied naval commanders breathed a sigh of relief that its development had not arrived earlier in the war.

It was one of a number of significant developments in German U-boat technology which were about to come on stream as the war reached the final stages. By then, it was already clear that submarine technology was on the threshold of dramatic developments. The Allies made a dash for the German U-boats and their accompanying manuals, which were to provide some inspiration for a new breed of submarines built on both sides of the Iron Curtain after it had slammed down at the start of the Cold War.

The Walther engine, used in four U-boats built before the war ended, was put under the microscope. It was a steam-generating power plant of a kind that had eluded British and American submarine engineers for years. It worked on the principle that hydrogen peroxide, when passed over a catalyst, produced oxygen and water. The oxygen and water were fed into a combustion chamber, sprayed with fuel, and the resulting mixture generated steam that powered a turbine. Walther had been working on the engine for four years, but because of its revolutionary principles there was little chance of proving it at the time of increasing pressure on the U-boat command.

One of the Walther boats commissioned in March 1945 was scuttled at the German port of Cuxhaven on 5 May 1945. The British had already learned of its existence and raised the boat by the end of the month. It was recovered, towed back to England, rebuilt and commissioned as *Meteorite* for evaluation. By then, the two big powers – picking over the bones of German technology – were on the threshold of a new era. Nuclear power was in sight.

BELOW: **Sections of the Type XXI, the first real combat submarines whose technology the Allies could not wait to get their hands on. Perhaps this type could have won the Battle for the Atlantic, had she arrived earlier.**

# Unfinished business: surviving a sinking

With so much loss of life due to submarines lost during the war, a special committee formed in 1939 after the *Thetis* disaster was reactivated under the chairmanship of Rear-Admiral P. Ruck-Keene, and was able to draw on a vast wealth of experience. He was given a large team of experts to assist him, and they began by taking detailed statements from survivors of 32 submarine incidents involving British, American, Norwegian and German vessels. All past methods of escape were examined, and new ideas were brought forward, extended training programmes drawn up and a better system for rescue response recommended and acted upon.

The immediate outcome of the early testimony was to improve escape training for crews. For this a 30m/98ft 5in water-filled tower, 5.5m/18ft in diameter, was built at the submarine school at Fort Blockhouse, although it was not completed until 1952. The rescue alerts for submarines in distress, with particular attention to reaction times, were also overhauled as part of the two systems called Subsmash and Subsunk. Some of the modifications to submarine safety and rescue procedures were thus to be tested sooner than anyone could possibly imagine. Less than five years after the end of the war, Britain suffered two major submarine disasters.

On January 12, 1950, the World War II submarine *Truculent* spent the day at sea off the Thames Estuary carrying out trials, following a long refit in Chatham dockyard. On board was her crew of 59, plus 18 civilian dockyard technicians who were to check the final performance before officially handing her back to the Royal Navy. As *Truculent* made her way back to Sheerness, she was in collision with a tanker with ice-breaking bows, *Dvina,* on passage from Purfleet bound for Sweden with a cargo of paraffin. Sixty-seven men made a near-perfect escape from the sunken submarine and, apart from one or two problems with untrained technicians, the Davis gate saw everyone get to the surface alive. However, in the end, only 11 men were picked up alive. The remainder died of the cold or were swept away by the tide.

At the subsequent inquiry, it was observed that had the men remained inside the submarine for 20 or 30 minutes longer, the rescue ships sent under the Subsunk alert would have arrived over *Truculent* to save them. By the same token, it was also pointed out that because of the flooding and the closure of all but the after-compartments of the submarine, the 67 men on board were already beginning to feel the effects of carbon dioxide poisoning when the first officer made the decision to get them to the surface. He had little choice. If he had delayed, they might have been in no state to make the attempt, as had been the case in so many previous incidents examined in recent times. *Truculent* was salvaged on March 14, 1950, and

**ABOVE LEFT: The salvage operation to raise HMS *Truculent*, which sank after a collision. Even though almost all aboard escaped, only 11 survived on the surface. ABOVE: U-boat crews were well advanced in their training to escape a catastrophe. LEFT: HMS *Affray*, which was lost on a patrol exercise in 1951. The search for her kept the British nation agonizingly enthralled for days.**

beached at Cheney Spit. The wreck was moved inshore the following day, where ten bodies were recovered. She was refloated on March 23, 1950, and towed into Sheerness Dockyard. A subsequent inquiry attributed blame both to *Truculent* and to a lesser extent *Dvina*. The loss led to the introduction of the "Truculent light", an extra steaming all-round white light on the bow of British submarines.

The *Truculent* disaster was comparatively fresh in the memories of submariners when a second post-war tragedy struck, with further implications. At 16:30 hours on the afternoon of Monday, April 16, 1951, the six-year-old A-class submarine *Affray* left Portsmouth for a war patrol exercise with an unusually large number of men on board. Under the command of Lt John Blackburn, she carried his team of four officers and 55 ratings. There were also 23 young sub-lieutenants in training, who came along to observe a simulated war patrol, and a party of four marines. Blackburn's orders were to proceed down the Channel to the Western Approaches, carry out dummy attacks for three days to give experience to the junior officers and then land the marines at a bay on the Cornish coast.

At 20:56 hours, Blackburn signalled from the Isle of Wight that he intended to dive and proceed westwards up the Channel. He would surface at 08:00 hours the next morning. No surfacing signal was received, and at 09:00 hours on the 17th a Subsunk operation was launched. By midday, forty ships, eight submarines, various aircraft from the RAF and Fleet Air Arm and a flotilla of US destroyers visiting Plymouth were all engaged in the search for *Affray*.

But after 46 hours, the massive search operation was stood down. The task of locating her was handed over to a smaller team, consisting of four frigates and three minesweepers equipped with all the special electronics available for underwater location. It was an epic search lasting 59 days, covering 37,013km/23,000 miles and involving the investigation of 161 shipwrecks in the search area before the submarine was found. Navy divers discovered the snort mast had broken off, but that probably occurred when the sub hit the bottom. Suggestions of a battery explosion were also evaluated, but in truth the cause of the sinking could not be ascertained. What was brought home once again, however, was that dramatic new escape systems were needed. The *Affray* was the last British submarine to be lost with all hands.

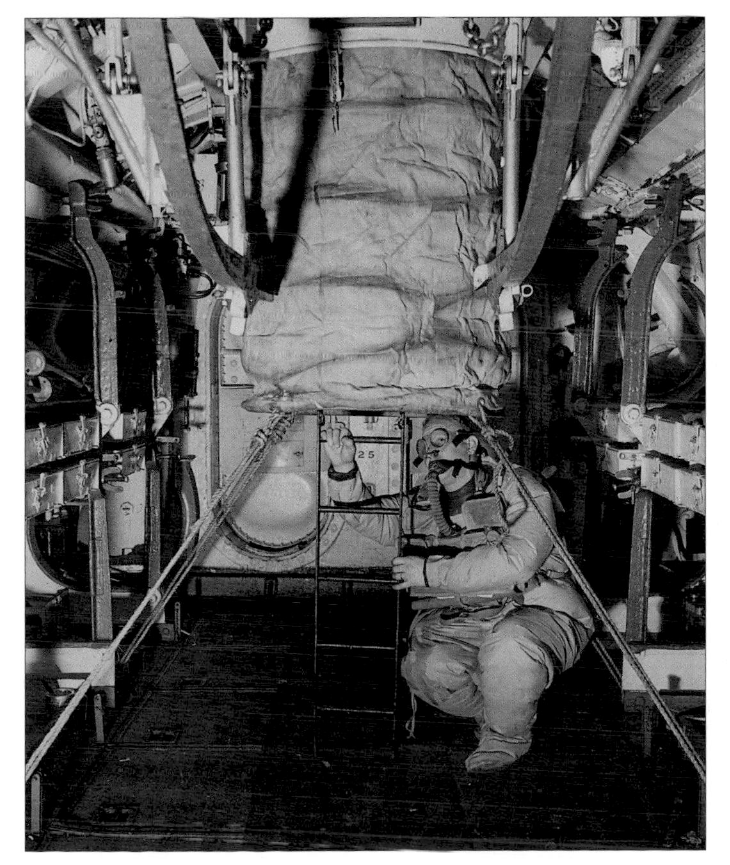

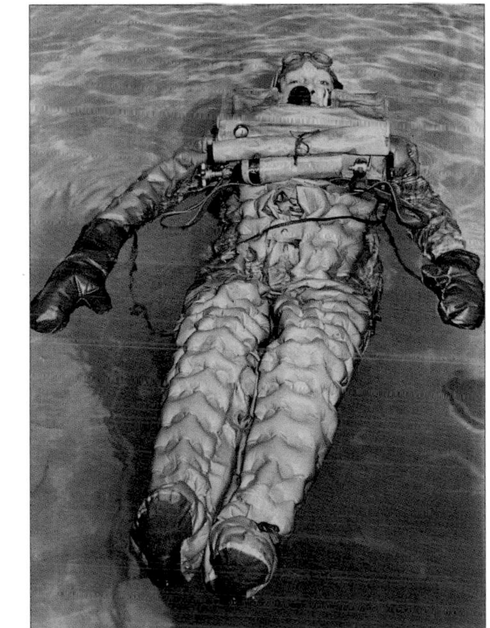

**ABOVE: Immediately after the war, when so many recent experiences could be recalled, the safety issues for the escape and rescue of personnel from a stricken submarine were the subject of numerous studies and experiments, including a large tank for training purposes with breathing apparatus. LEFT: Other training methods and equipment were tried, but survival remained something of a haphazard affair. ABOVE: Immersion suits and life jackets were also tested, although their suitability for escaping from a submarine – as opposed to surviving an accident or the sinking of a surface ship – proved to require many more years of trial and error before satisfactory solutions to catastrophe beneath the waves were found. Meanwhile, the accidents and tragedies continued to occur.**

# The Cold War: teardrops and Guppies

After World War II, the beginning of an undersea warfare revolution brought the United States and the Soviet Union into a confrontation in which they were the only two serious players – and it was the submarine forces of the two nations on which the naval developments were stridently focused. Building on the advanced German XXI submarines – the United States, Russia and Britain took possession of one apiece – the US Navy anticipated submarines of the future going deeper, staying there longer, and moving much faster. However, they weren't alone in these endeavours, of course, and naval and civilian advisers warned that in exploiting the XXI type, the Soviet Union presented "the most potent post-war naval threat to the United States".

It was known that the Soviets, still repairing the devastating effects of the war, lagged behind the Americans. They publicly boasted a potential force of 300 Soviet Type XXI equivalents by 1950, but it was not until 1949 that the first post-war Soviet submarine designs went to sea. Two classes were deployed: the Whiskey and the Zulu. The Zulu was a true Type XXI, equipped with a snorkel, capable of 16 knots submerged. But only 21 Zulus were commissioned between 1949 and 1958. During the same period, 236 Whiskeys were commissioned. They were smaller, shorter range boats, designed more for European littoral operations.

Technology was still in its infancy and intelligence gathering tended to be of the manual kind. US submariners were on round-the-clock front-line positions to gather clues as to Russian advances. In August 1949, for instance, submarines *Cochino* and *Tusk* were deployed to the waters off Norway 161km/100 miles from the Soviet Northern Fleet's bases at Murmansk and Polyarnyy to learn what they could about Soviet missile testing. Without warning, batteries in *Cochino* exploded and badly burned one officer. Fire and noxious gases released

TOP: **Initially, the Soviet Union lagged behind the West. Their largest class of new boats was the Whiskey class, of which 236 were built between 1949 and 1958, all of conventional shape.** ABOVE: **USS *Albacore* changed everything. Her design became the model for the teardrop hull adopted in all modern submarines across the world, and was critical in the race for the development of superior submarines.**

by this and subsequent blasts spread throughout the submarine, endangering the entire crew. Two *Cochino* men, trying to bring help from nearby *Tusk* while maintaining radio silence, were pitched into the bitterly cold water when their rubber boat overturned. Without hesitation, men from *Tusk* jumped in to help rescue their fellow sailors. Several men drowned in the attempt and their bodies drifted silently off into the northern waters. Finally, after heroic efforts by everyone, the surviving crewmen of both submarines gathered safely on board *Tusk*, and as they prepared to set sail for home, the *Cochino*, gutted by fire and explosion, vanished beneath the waves.

This was one of the last occasions that such primitive cloak and dagger work was undertaken. The Guppies made sure of that. This was the name given to the boats produced for the sole purpose of "Greater Underwater Propulsive Power". They were developed in the United States, and followed by the

ABOVE: USS *Cochino*, a *Balao*-class submarine, commissioned in 1946, was transformed into one of the world's most modern submarines with the Guppy conversion. She sank after a severe battery explosion in stormy seas off northern Norway, though the crew was rescued by USS *Tusk*. ABOVE RIGHT: USS *Pickerel*, a *Tench*-class submarine, was another boat converted to Guppy, and in 1950 she made a record 8,369km/5,200 miles submerged trans-Pacific voyage. RIGHT: USS *Blueback*, commissioned in 1959, was the first US submarine built specifically to the Guppy configuration (as opposed to a conversion) and was also the last non-nuclear boat to take her place in the US Navy. Key features included removing the deck guns, streamlining the outer hull, replacing the conning tower with a sail, installing more air conditioning and a snorkel, and doubling the battery capacity. She can be seen in the Sean Connery movie *The Hunt for Red October* in the role of USS *Dallas*.

Soviets both using features borrowed from the XXI. The star elements of the Guppy design consisted of the removal of the deck guns, streamlining of the outer hull, installing new propellers for submerged operations and – most distinctively of all – replacing the conning tower with a sail. Initially, existing boats were converted to take the new developments, and instantly underwater speeds were almost doubled. The first Guppy sent to the Far East, *Pickerel* made the entire trip back from Hong Kong to Pearl Harbor on snorkel – 8,359km/ 5,194 miles in 21 days submerged, a record that was held until 1958. In the new boats that were to get these innovations as standard, the Tang class first commissioned in 1951 led the field in this development. The *Tang* was the first truly new post-war construction that represented the first step towards greater speed and endurance below the surface.

There was one other innovation that was to change the look of submarines forever. American researchers and designers had been drawing new shapes on their board since 1944. In 1950, the one that grabbed the eye of everyone was an attractive hydrodynamic hull shape that became known as the teardrop.

A prototype was built and extensively tested in wind tunnels, subsequently leading to a $20 million development budget out of which came the *Albacore*: the first teardrop submarine, fast and whale-like with a sail distinctively positioned a third of the overall length from the bow.

She was commissioned in December 1953, and initial trials showed that she was faster submerged than the designers had anticipated. The attractive design set the standard, and was adopted in due course by most navies of the world.

BELOW: USS *Tang* was the first of the American new breed to be designed post-war and as such was the first of six that incorporated the Guppy (snorkel) configuration into the design, as well as being test platforms for the first nuclear submarines.

LEFT: **USS *Nautilus*, the sixth American ship to carry that name, was the world's first operational nuclear-powered submarine. Heading south for her shakedown cruise, she travelled 2,100km/1,300 miles from New London to San Juan, Puerto Rico, in less than 90 hours, which was the longest ever submerged cruise by a submarine and at the highest sustained speed ever recorded. Later, she became the first vessel to complete a submerged transit across the North Pole. By February 4, 1957, *Nautilus* had logged 111,120km/ 60,000 nautical miles, matching the endurance of the fictional *Nautilus* in Jules Verne's *Twenty Thousand Leagues under the Sea*.** INSET: **The badge of SSN-571, the first nuclear-powered attack submarine.**

# *Nautilus*, the first true submarine

And so to the vessel that Jules Verne predicted in his 1870 novel, *Twenty Thousand Leagues under the Sea*: a true submarine that could travel the globe without ever surfacing, a fictional construction that bore many similarities to the reality that finally emerged in 1955. Even the name was the same: *Nautilus*. It all began in the mid-1940s, when Captain Hyman Rickover, a brilliant naval engineer and experienced submariner, was burning the midnight oil, working on his own theories for such a boat. Without any great enthusiasm from more senior naval figures, Dr Ross Gunn of the US Naval Research Laboratory and Dr Philip Abelson of the Carnegie Institute joined Hyman in 1948. Both had been working on similar themes that would bring about a revolution in submarine propulsion on similar lines to the Walther boat, and indeed their eventual product bore a number of similarities. The difference was that unlike Walther who used highly concentrated hydrogen peroxide, Gunn proposed the use of nuclear fission.

In the simplest terms, a reactor would generate steam, which would drive the turbine. For the next two years they struggled with the theory of a nuclear-powered submarine while at the same time fighting a relentless rearguard action against formidable opponents. Early in 1950, Rickover's team produced designs for a land-based nuclear reactor as a prototype for one that could power a substantial submarine. They demonstrated how a single power plant could provide the propulsion for unlimited surface and submerged travel. A small quantity of enriched uranium would produce enough power to run for years, and the nuclear submarine would be able to operate at high speed, completely and indefinitely submerged and restricted only by human endurance.

ABOVE: **All eyes on the monitors as *Nautilus* travels under the North Pole. The main Control Room was located directly below the Attack Centre, accommodating the instruments and controls for diving, surfacing and steering the ship. The Diving Officer of the Watch was in charge of this space and received orders for depth, course and speed from the Officer of the Deck in the Attack Centre.**

Rickover drew the comparison against even the most modern conventional diesel-electric submarines, in which the submerged approach to a target had to be made at a very low speed of no more than three knots, to avoid wasting battery power, and whose ability to stay submerged was limited to the state of on-board batteries. As every submarine commander well knew, this was critical in any attack situation because sufficient power always had to be husbanded in case of counter-attack and the possibility of having to sit on the bottom

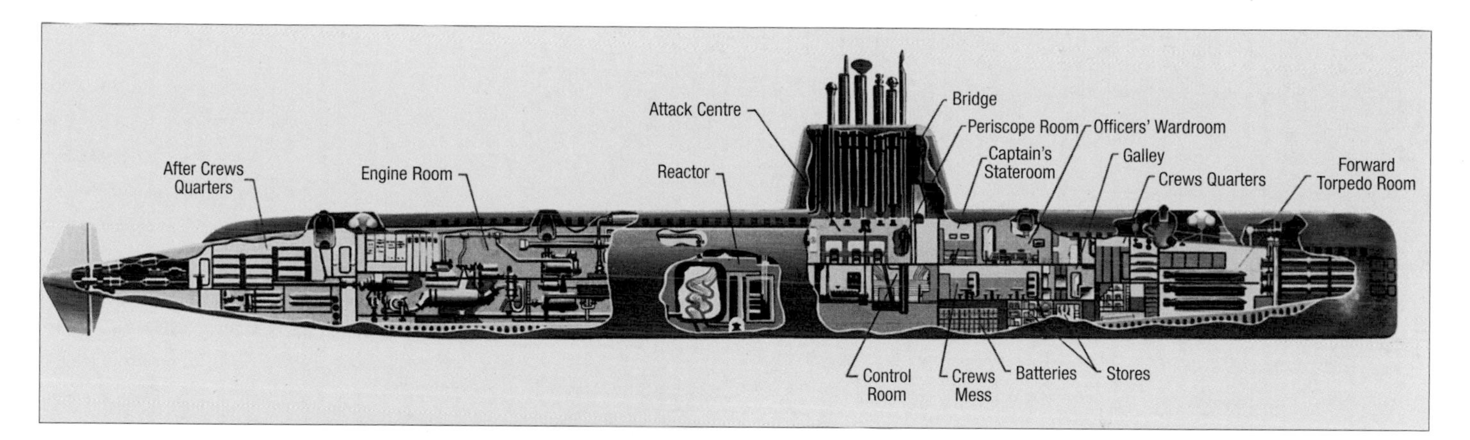

ABOVE: **Apart from the historic nature of the submarine's development, great emphasis was placed on interior design, as is evident from this compartmentalized view of the various operational stations.** LEFT: **A fine starboard view of SSN-571 *Nautilus* at sea.** BELOW: **A view of the Torpedo Room, which held six tubes, initially for MK 14 torpedoes as displayed above, as well as stowage racks for 24 torpedoes. Directly aft of the Torpedo Room was a small berthing area with 10 bunks, toilets, shower and sinks for the weapons crew.**

until the danger had passed. The need to conserve battery power had always meant that submarines had to be cautious about engaging fast surface warships. Nuclear submarines, said Rickover, would end all those problems overnight. They would be fast, they could travel at top speed without ever having to surface and they could remain submerged for great distances. The size of the nuclear reactor and its shielding meant that the submarine had to be 97m/318ft 3in long and displace 3,539 tonnes/3,483 tons on the surface, which was more or less the size of a modern cruiser.

Many were unconvinced by these claims; some were even scared by them. Even so, the US Congress approved a $30 million budget to build the world's first nuclear-powered submarine, and her keel was laid down by President Harry Truman at the Electric Boat Company shipyard – the same company that had produced the first Hollands at the turn of the century – in Groton, Connecticut, on June 14, 1952. On the morning of January 17, 1955, the first commanding officer, Commander Eugene P. Wilkinson, ordered all lines cast off and signalled the historic message: "Under way on nuclear power."

Very soon, *Nautilus* began to shatter all speed and distance records, and in August 1958, then under Commander Bill Anderson and with 116 men on board, she left Pearl Harbor under top secret orders to commence Operation Sunshine – the first crossing of the North Pole by a ship, a mission that many said was impossible. The achievement became known to the world when, on August 3, *Nautilus* reached the geographic North Pole 90 degrees north and Anderson signalled a message that would be echoed by the moon landing team a few years later, he said: "For the world, our country and the navy – the North Pole."

These developments came at a time of mounting East-West tension and, we now know, were to have a profound effect on the race for nuclear domination by the superpowers.

# *Skipjack* sets the standard

By 1950, Americans had nuclear power in a relatively conventionally shaped boat, *Nautilus*, and the teardrop design of *Albacore*, which had proved exceptionally fast, even with conventional power. Next, the two technologies were merged to create the Skipjack class, combining the endurance of nuclear propulsion with the streamlined hull that at a stroke would produce the world's fastest submarine, said to be capable of an incredible 30 knots dived. The American team had also created the design base for all future US submarines.

Five Skipjacks were ordered, and were subsequently commissioned during the coming decade. They were essentially attack submarines in support of the American anti-submarine warfare programme, codenamed SOS-US, which came into effect after US intelligence surmised that any atom-bomb attack on the United States would most likely be carried by submarines.

This was extrapolated from the fact that both sides were already working on the development of ballistic missiles, again based initially on the German technology of the V1 and V2 rockets that had caused so much devastation in wartime London. From that one assessment, the whole development of attack submarines subsequently began to take on a new complexion, as indeed did America's own anti-submarine protection. From the late 1950s, a detection system revealing the advance of any submarine towards the American east coast was installed along the entire coastline, from north to south, and later extended to cover the west coast and Hawaii.

*Skipjack* was also unique in that it was the first nuclear submarine with a single shaft. Placement of the bow planes on the sail greatly reduced flow noise at the bow-mounted sonar. The deep-diving and high speed capabilities of the class were the result a new reactor design, the S5Ws, that became the US Navy's standard until the Los Angeles class joined the fleet in the mid-1970s.

Further developments focused on reducing the noise levels of the Skipjacks, which were easy to detect even at extreme range. For this, the designers turned to the smaller *Tullibee*, which had been designed specifically as an Anti-Submarine Warfare (ASW) weapon. These boats were noted in US submarine history as the quietest nuclear submarine with turbo-electric drive, the first with a sonar suite that included a low-frequency passive array for long-range detection and

LEFT: **USS *Shark*, a *Skipjack*-class submarine commissioned in 1961, later joined the Sixth Fleet in the Mediterranean, the first such deployment for a nuclear submarine. In 1964, she was awarded a commendation for "achieving results of great importance in the field of anti-submarine warfare". She was decommissioned in 1986.**

ABOVE: **The Soviet *Hotel*-class ballistic missile submarine was put into service in 1959, designed to carry the D-2 launch system. This version, from the 1969 era, with a NATO reporting name of Hotel III, had been lengthened to test R-29 missiles.**

LEFT: **USS *Tullibee*, commissioned in 1960, was something of a test platform, with innovative designs. She was nicknamed the "boat of firsts" – the first designed for anti-submarine warfare, the first to be equipped with a new series of sonar equipment, the first with torpedo tubes amidships and the first to use turbo-electric propulsion.**
ABOVE: **The Russian November class was also a boat of firsts – the first Soviet nuclear-powered submarine in 1958, the first to traverse the North Pole (four years after the Americans) and the first to carry ballistic weapons when the class was modified.**

LEFT: **USS *George Washington* was already under construction when the nuclear arms race opened up, and was cut in two for the installation of a ballistic missile compartment, to become America's first ballistic boomer.**
ABOVE: **USS *Scamp*, a *Skipjack*-class submarine, was among the early nuclear boats later converted to receive SubSafe, whereby every component and every action is intensively managed. The system added significant cost, but no submarine certified by SubSafe has ever been lost.**

a spherical array for approach and attack, and the first submarine with torpedo tubes amidships. This step was a further milestone in the history of nuclear boats, which signalled the development of the multipurpose SSN that combined the speed of *Skipjack* and the ASW capability of *Tullibee* into one boat. This became the *Thresher* from which all future US submarine classes derived. The acoustic advantages were hailed as dramatic, and the new boats were in a class of their own when compared with the fleet that the Russians were in the process of building. These were colloquially known as the Soviet HENS, standing for the initials of the Red Navy's principal classes, Hotel (SSBN), Echo (SSGN) and November (SSN), which began deploying alongside the Threshers in the early 1960s as these great monsters jostled for position and supremacy.

However, there was yet to be one more crucial development based on the *Skipjack* model. Since the end of the war, both nations had been experimenting with submarine-launched ballistic missiles, and *Skipjack* proved to be an ideal platform for that purpose. It was the beginning of the true era of potential nuclear annihilation launched from beneath the waves. Unlike easily targeted land-based missiles, submarines carrying strategic ballistic missiles (SSBNs) were constantly on the move, lurking deep in the ocean, with almost unlimited staying power. They were regarded as being capable of reaching almost any target in the world, and after the successful trials of the Polaris A1, *Scorpion*, an SSN of the Skipjack class, was sliced in two for a 39.6m/130ft missile section to be inserted to create the first nuclear-powered ballistic missile carrier. *Scorpion* was renamed *George Washington*, thereby establishing a new class of submarine. This giant weighed in at 5,900 tonnes/5,807 tons, was 116.4m/382ft long and carried 16 Polaris missiles with a range of 2,222km/1,200 nautical miles. They were to be matched and bettered, at least in terms of size, by the Soviet Union's 8,128-tonne/8,000-ton Yankee class, which eventually packed in an armoury of 16 SS-N-6 missiles with a range of 2,408km/1,300 nautical miles.

LEFT: **Britain's first nuclear powered submarine, *Dreadnought*, whose endurance and versatility demonstrated the limitations of the rest of Britain's submarine fleet, especially when she travelled from Rosyth to Singapore and back in 1969. Later, she became the first British submarine to surface at the North Pole.** BELOW: **The modernity of the control room was, at the time, an eye-opening experience for all newcomers to an era of nuclear-powered travel, which of course also required a new elite at all levels of submarine manpower.**

# Life aboard the *Dreadnought*

By 1950, Britain, once the greatest sea power in the world, was still suffering badly from the cash constraints imposed by the war and could not even begin to contemplate anything more than a tag-along role to the United States in the two-horse race for supremacy in submarine warfare. Her shipyards were still very capable of producing world-class boats, of course, but for the present her submarine force was still anchored in wartime technology that would barely raise a blip on the modern radar screen. Her only nuclear deterrents were the RAF's V-bomber force and land-based missiles strung down the east coast pointing towards the Soviet Union. However, Lord Louis Mountbatten, as First Sea Lord and later as Chief of Defence Staff, was an ardent campaigner for Britain's own nuclear submarines and eventually persuaded the British government to provide the finance to begin the process of turning the nation's submarine force into an all-nuclear complement, a process that would take several years.

It began with the building of Britain's first nuclear-powered submarine with a power plant bought from the Americans. She was to be christened *Dreadnought*, the ninth ship in British naval history to carry that most famous name, and in a subliminal way this perhaps demonstrated that the Admiralty had finally acknowledged that the submarine had graduated to capital ship. She was launched by the Queen at Barrow-in-Furness on Trafalgar Day, October 21,1960, by which time, incidentally, the United States had stopped building conventional submarines altogether.

*Dreadnought* was a proud bearer of a bit of British face-saving in a world where submarine power was assuming astronomical proportions. She had a surface displacement of 3,500 tonnes/3,445 tons and 4,000 tonnes/3,937 tons submerged. She was 78m/256ft long, had a beam of 10m/32ft 10in, could travel effortlessly at 28 knots submerged and had a crew of 88. She was classed as a hunter-killer fleet submarine, armed with six conventional bow torpedoes. She was to be the prototype for the creation of Britain's nuclear submarine force, which from now on would be built entirely by British scientists, engineers and shipbuilders, with more planned for the immediate future, to form the Valiant class. The price the First

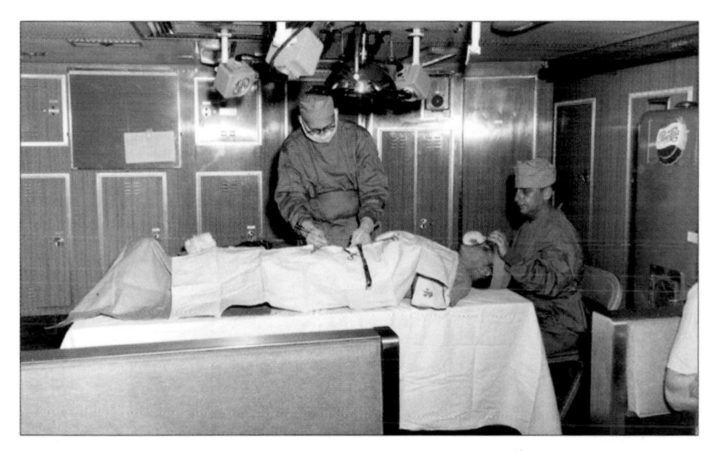

ABOVE: **Nuclear propulsion also presented other considerations, given the greater length of time spent submerged. The medics on the 1969-built USS *Andrew* had to perform an emergency operation on a makeshift table in the mess room, although by then newer ships were already being fitted with full medical suites.** BELOW: **An image demonstrating the sleek lines of HMS *Dreadnought*, which rightly instilled pride among the British submarine community.**

ABOVE: **Catering was also to become a major new consideration for submarines of the nuclear age with enough chefs to prepare meals for several sittings in pleasant, clean and hygienic surroundings, carrying enough stores for extended journeys at sea, and with a varied and healthy diet of fine quality fruit, vegetables and meat.** BELOW: **The mess halls on the nuclear submarines also took on the air of a relaxing, spacious restaurant with modern accoutrements.**

Sea Lord had to pay for that agreement from the government was a halt on the building of further diesel-electric submarines, which in turn left Britain very short of boats a decade later.

*Dreadnought* delivered the long-awaited revolution to the British submarine force. She was new in every regard, handled by telemotor controls using a joystick and a dazzling array of dials that made up the elaborate instrument panel. For the crew going aboard for the first time, it was a jaw-dropping experience, with accommodation to a standard they had never seen before, even in surface vessels. Apart from an efficient air-conditioning and purification system, there were numerous luxuries, such as showers, washing facilities, a laundry, and a large galley equipped for serving decent meals. There were also separate messes for junior and senior ratings, well-furnished recreational spaces and a range of leisure facilities, such as a cinema, a library and other features to relieve the monotony of long periods submerged.

By then, the British government had approved the building of *Valiant*-class boats, for which *Dreadnought* was the prototype. Five were to be laid down for construction between 1962 and 1970, and with that approval came a major operation in recruitment and retraining to prepare the crews and shore bases to take control of these revolutionary boats. Nor was it simply a question of training. As had been discovered in the United States, the unique qualities of life aboard a nuclear submarine required commanders of outstanding leadership skills and crews capable of withstanding the particular pressures that came with long periods of life beneath the waves. There was no precedent to the experience. The whole point of nuclear boats was that they would submerge as soon as they left their base and, if necessary, remain submerged for the duration of their patrol. From the very beginning, crews might go for weeks without seeing daylight or having any contact whatsoever with the outside world. Most on board would not have the slightest inclination of their whereabouts, the time zone they were in or even whether it was night or day.

Two, three or four month patrols submerged became commonplace, and the aspects of crew welfare necessarily became a priority in both selection and training. The fact of life that had faced all submariners since the beginning – the confined and restricted nature of travelling in a boat tight for space and packed with dangerous gear – was no less pronounced in the comparatively palatial surroundings of the nuclear boats. If anything, they were heightened by the longer periods of time spent submerged in what was still a confining, if not confined, space. Greater studies would follow, but from the early American experience, repeated in Britain, it was immediately evident that prospective crews needed careful screening for temperament, intelligence and fitness. Selection and training of commanders, officers and senior ratings in turn became perhaps the most crucial elements of all. In their hands rested the responsibility for all elements, human and mechanical, in these highly sophisticated war machines.

# Directory of Submarines

In 1900, the Royal Navy did not possess a single submarine. In fact, Britain was the only major maritime power not to have even the beginnings of a submarine flotilla. The United States Navy was trialling her own *Holland I*, and the Royal Navy followed the US lead in 1901. Thereafter it became a race to produce boats of ever-increasing endurance and sophistication.

By the outbreak of World War I, the unimaginable was already happening: submarines could actually take on the mighty surface ships – and win. Between the wars, design emphasis switched to diesel-electric boats, and considerable advances had been made by the onset of World War II. With the outbreak of hostilities, Germany turned to her submarines to take the war to the Allies at sea, using their boats as submersible destroyers.

Britain relied on her new O, S, T and U-class boats, but was forced to begin with older stock. The Japanese had begun producing giant boats, while the Americans went into mass production with a number of major classes, such as Gato, Balao and Tench, that were to become the workhorses of the US submarine fleet.

LEFT: **British submarines of the 10th Flotilla based at Lazaretto, Malta, included *Upright* (second from right), which sank Mussolini's cruiser *Armando Diaz* on February 25, 1941.**

LEFT: **With no room on top, the crowded state of the early submarines meant a wet start to the day for the inexperienced, and it was even more crowded down below.** BELOW: **The classic shape of the *Holland* boat, repeated in magnified proportions more than half a century later.**

# *Holland*

After much deliberation, submarines formally became part of the US Navy's inventory in 1899 when a series of trials of the *Holland* convinced the Navy Board that they should go ahead. It had been a long and difficult journey for her creator, J.P. Holland, who had won a competition staged by the Navy 11 years earlier to build a torpedo-firing submarine that could "travel submerged for two hours at eight knots at depths up to 150 feet".

Although he won the competition, Holland did not receive a contract to build his boat, and it took several more unsuccessful attempts that saw him slide into near bankruptcy before the US Navy finally came up with a $165,000 contract for *Holland VI*.

Launched on May 17, 1897, at Elizabethport, New Jersey, it was equipped with a 37.3kW/50hp petrol engine for surface propulsion, for which a range of 1,609km/1,000 miles was claimed. The US Navy was impressed during trials when the boat managed to get within 91.4m/100 yards of the US battleship *Kearsage*, whereupon a light was flashed to demonstrate that the ship could have been sunk with the submarine's 450mm/17.7in Whitehead torpedoes.

The *Holland* was built under her designer's supervision at the Crescent Shipyard at Elizabethport, New Jersey, and was noted for her sleek lines in the likeness of a porpoise. In reality, however, she travelled low in the water when on the surface, making it precarious for her crew to work on deck, especially when operating the Zalinski dynamite gun mounted on the bow, which was subsequently removed. The Navy contracted for the building of five boats, although there were serious doubts as to their viability among those who still had to be convinced of the usefulness of submarines. Not least among these were some of the sailors who were destined to man the new contraptions. The boat's stability on the surface left much to be desired and the internal arrangements were spartan and dangerous.

Inside the boat, the slightest malfunction of the exhaust system from the petrol-driven engine could kill the entire crew by carbon monoxide poisoning. Even a minor fuel leak might cause an explosion, as could battery problems, which plagued submarines for years hence. The *Holland* did not see any serious service, and was used mainly for training as the first new boats of an improved Holland design came on stream.

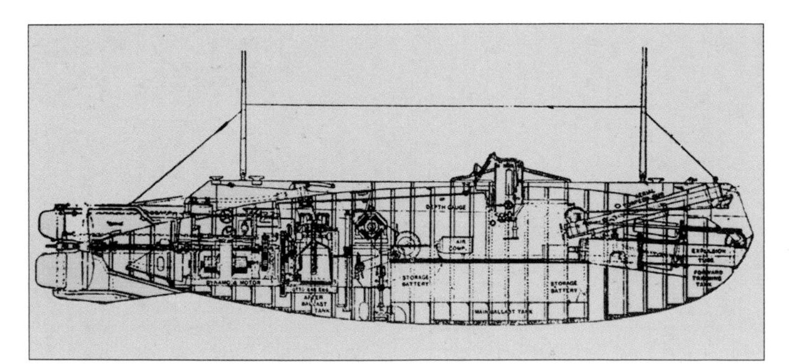

LEFT: **The original *Holland* drawings for his boat, more or less adhered to by the US Navy designers.**

## *Holland*

**Displacement:** 62.98 tonnes/62 tons (surfaced); 72.83 tonnes/71.7 tons (submerged)
**Length:** 16.18m/53ft
**Beam:** 3.13m/10ft 3in
**Armament:** 1 x 457mm/18in bow tube, 2 x torpedoes
**Propulsion:** 33.6kW/45hp petrol; 41kW/55hp electric
**Speed:** 8.0 knots (surfaced); 5.0 knots (dived)
**Complement:** 9 men

LEFT: **USS *Porpoise* (which became *A6*) and *Shark* (*A7*) were the last two boats of the A class. Both were eventually loaded aboard a collier and transported to the Philippines for local patrols.**
BELOW: **The safest way to keep watch, since periscopes were yet to be invented.**

# A class

The seven boats of this class entered service in the first three years of the new century, all based upon an enlarged version of the *Holland I*, although confusingly the first boat in the class, originally named *Plunger*, was classified as *A1* on the creation of the class. However when all US submarines were given the SS classification in 1920, she became *SS2*, the first Holland to enter service becoming *SS1*.

The *Plunger* and the rest of the A class saw considerable service in those early years, and despite all problems concerning human habitation inside the boats, remained in service well into the second decade of the century. Three were reconditioned for coastal work and training in the early years of World War I. *Plunger* also went into history as being the first submarine to have a President of the United States aboard. She was conducting trials at Oyster Bay, close to the home of Theodore Roosevelt, in the summer of 1905. The crew had been given advance warning, giving them time to paint the outside of the boat before the president stepped aboard on August 22, 1905.

He spent almost three hours on board, and was given a full demonstration and experienced a number of dives, an event that of course captured the interest of the media and public, many of whom were still somewhat sceptical about the safety of these underwater creatures. Roosevelt took note of the reaction, and later wrote, "I went down in it chiefly because I did not like to have the officers and enlisted men think I wanted them to try things I was reluctant to try myself. I believe a good deal can be done with these submarines, although there is always the danger of people getting carried away with the idea and thinking that they can be of more use than they possibly could be".

Later, *Plunger* hosted another figure of subsequent fame: Chester W. Nimitz became her skipper in May 1909, the first command of the man who would rise to become Fleet Admiral. Although at the time Nimitz said he considered submarines "a cross between a Jules Verne fantasy and a humpbacked whale", he later became one of America's most distinguished naval figures. The experience on board *Plunger* and others of the early classes, however, was unforgettable. A Navy Medical Officer's report at the time was scathing about conditions on the boats, with the crew sleeping, cooking, eating, and answering the calls of nature in such a confined space. He said even in moderate seas, the boat rolled and pitched so that practically the whole crew was seasick. Food had to be packed in crates and the cooked meats soon spoiled and, coupled with the use of the open toilet, the air was foul.

RIGHT: **Assigned to the Naval Torpedo Station at Newport, Rhode Island, *Plunger* operated locally for the first two years.**

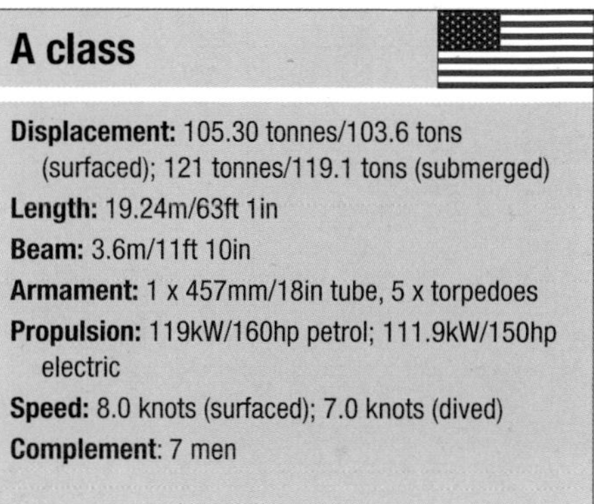

## A class

**Displacement:** 105.30 tonnes/103.6 tons (surfaced); 121 tonnes/119.1 tons (submerged)
**Length:** 19.24m/63ft 1in
**Beam:** 3.6m/11ft 10in
**Armament:** 1 x 457mm/18in tube, 5 x torpedoes
**Propulsion:** 119kW/160hp petrol; 111.9kW/150hp electric
**Speed:** 8.0 knots (surfaced); 7.0 knots (dived)
**Complement:** 7 men

# B class

Three boats of this class, *Viper*, *Cuttlefish* and *Tarantula* (*BI*, *B2* and *B3*), commissioned in 1907, saw the final stage of development in the Holland single-screw design and introduced a more extensive superstructure for sea keeping. On the original plans, *Viper* had only the single periscope let into the conning tower; however a second periscope was added later. Air compressors and the main bilge pumps were driven from the main shaft via noisy clutches and gears. The boats were altogether more substantial in appearance, and length, stability and conditions aboard were better, but still very cramped. A reload torpedo in the B class, for example, left little space in the hull. Even so, this class saw considerable service in the Atlantic Torpedo Boat Fleet and in April 1914 was posted for an assignment in distant waters.

*B1* was towed to Norfolk to be loaded on board USS *Hector* to be carried to the Philippine Islands, arriving at Luzon in March 1915 to be recommissioned into the Submarine Division 1, Torpedo Flotilla, Asiatic Fleet. She remained in the Philippines throughout the rest of her working life before being decommissioned in 1921. *B2* and *B3* had a similar history to the first-born of the class.

*B3*, for example, operated along the Atlantic coast with the 1st and 2nd Submarine Flotillas on training and experimental exercises until going into reserve at Charleston Navy Yard on November 6, 1909.

She came back into service when she was recommissioned on April 15, 1910, and served with the Atlantic Torpedo Fleet until assigned to the Reserve Torpedo Group on May 9, 1911, and placed out of commission again in December 1912. Two days later, she was towed to Norfolk and loaded on board *Ajax* for transfer to the Asiatic Station. Arriving at Cavite, Philippine Islands on April 30, 1913, *B3* was launched from *Ajax* on May 12. She remained in the Philippines, where she served with Submarine Division 4, Torpedo Flotilla, Asiatic Fleet until she was decommissioned in July 1921, and subsequently used as a target.

TOP: **Beginning to look like a workable boat, the B class was a considerable improvement in terms of stability but conditions inside were still primitive.**
ABOVE: **The shape and accoutrements to the deck along with additional length gave the B class a more definitive shape.**

LEFT: **For the first decade, US submarines were attached to the Torpedo Fleet, until finally they were accepted as permanent, and the name was changed to Submarine Flotillas.**

### B class

**Displacement:** 142 tonnes/140 tons (surfaced); 170 tonnes/167 tons (submerged)
**Length:** 25.15m/82ft 6in
**Beam:** 3.85m/12ft 8in
**Armament:** 2 x 457mm/18in tube, 4 x torpedoes
**Propulsion:** 186.4kW/250hp petrol; 44.7kW/60hp electric
**Speed:** 9.0 knots (surfaced); 8.0 knots (dived)
**Complement:** 10 men

LEFT AND ABOVE: **Although subs had been around for a decade or more, there were still many doubters in the hierarchy of the US Navy, but these boats of the C class, *Octopus* (*C1*) (LEFT) and *Tarpon* (*C3*) (ABOVE) at last turned a few heads.**

# C class

With the entry of *C1*, originally christened *Octopus*, US submarines were beginning to be taken seriously by naval aficionados. Even so, the boats of the equivalent British class were already much larger – almost one third longer and with 447kW/600hp engines, compared with the 373kW/500hp for the US C class. The debate in the United States over the true value of submarines in warfare was probably still weighted towards the doubters, but the new designs were beginning to change minds.

In that regard, *Octopus* had the distinction of introducing what was to become the recognizable stern design of the Electric Boat Company, and was the first boat designed by L.Y. Spear following the departure of J.P. Holland from the company. It was also the first of the US boats that came with an air-operated bell for underwater signalling, which was then fitted to all earlier submarines. By the time *C5* (*Snapper*) was commissioned in 1911, further improvements were under way and this boat was also used for experiments with radio, signalling apparatus, alternative batteries types and a number of other innovations which eventually became standard. With war clouds gathering in Europe, the C class was also used in the first major trials in operations with warships, and later with aircraft with the advent of naval aviation. Lt Chester Nimitz wrote an interesting observation about *C5*:

*"Her Craig gasoline engines were built in Jersey City by James Craig an extraordinarily wise and capable builder. Craig was a self-taught engineer who began as a draftsman in the Machinery Division of the New York Navy Yard and who started his Machine & Engine Works in Jersey City at a later date. C5's engines were excellent as were the Craig diesel engines he built for a subsequent submarine. These engines were designed and built by Craig and I have never forgotten his Foreword to the pamphlet of Operating Instructions which read briefly somewhat like this: 'No matter what the designer and the builder may have planned for these engines and no matter what the operator may try to do with them the Laws of Nature will prevail in the End'. How True!"* Five C class were built.

ABOVE: **C-class submarines in the Gatun Locks, Panama Canal, just prior to World War I. The submarines in the line-up were *C1*, *C2*, *C3*, *C4* and *C5*.**

## C class

**Displacement:** 234.25 tonnes/230.6 tons (surfaced); 270 tonnes/265.7 tons (submerged)
**Length:** 32.12m/105ft 5in
**Beam:** 4.25m/14ft
**Armament:** 1 x 457mm/18in forward tube, 5 x torpedoes
**Propulsion:** 372.8kW/500hp petrol; 111.9kW/150hp electric
**Speed:** 10.5 knots (surfaced); 9.0 knots (dived)
**Complement:** 15 men

LEFT: **Getting visibly larger, a starboard-side view of *Grayling (D2)*, circa 1914, thought to be in the harbour of Newport, Rhode Island.**
ABOVE: **The somewhat spartan control panel.**

# D class

The experiences among all submariner nations gradually drew the emphasis in design and development towards greater safety for crew members in the event of a catastrophe. In conjunction with US Navy engineers, designers at the Electric Boat Company came up with some ingenious techniques first tried with *D1*, originally launched in 1908. She became the first US boat to be compartmentally divided specifically for survivability. Each compartment was defined by closely positioned bulkheads so that the submarine (surfaced) could survive flooding.

In fact, the layout proved to be difficult in terms of the complicated internal access because of the closeness of the bulkheads. The US Navy ordered a hasty re-design for *D3* that actually reversed the overall safety ideal that had been the original aim. As a result she had only two bulkheads, one at the after end of the torpedo room and one at the fore-end of the engine room. By pushing out the bulkheads, space in the control room was greatly increased, but the resulting compartments were so large that the boat was unlikely to survive if one of them was flooded.

This re-adjustment was considered by some analysts to be the cause of a number of US submarine losses in the 1920s, especially after collisions.

There were other experimental features with *D3* that led to further development, including placing the two periscopes very close to each other, which made it possible to brace both against vibration, a very serious problem with these boats. The object of the forward periscope was to act as a ventilator. The original drawings for *D3* also showed an improved temporary bridge, with a portable canopy frame and external ladder, protected by a trunk. Three D-class submarines were built.

LEFT: ***Grayling*, moored prior to a Presidential Review of the fleet in North River, New York.**

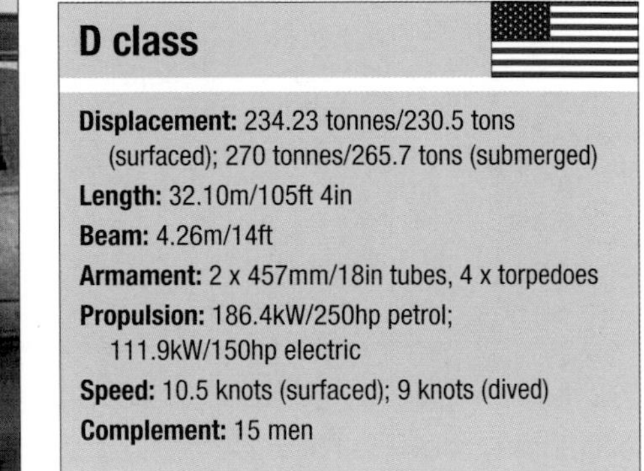

## D class

**Displacement:** 234.23 tonnes/230.5 tons (surfaced); 270 tonnes/265.7 tons (submerged)
**Length:** 32.10m/105ft 4in
**Beam:** 4.26m/14ft
**Armament:** 2 x 457mm/18in tubes, 4 x torpedoes
**Propulsion:** 186.4kW/250hp petrol; 111.9kW/150hp electric
**Speed:** 10.5 knots (surfaced); 9 knots (dived)
**Complement:** 15 men

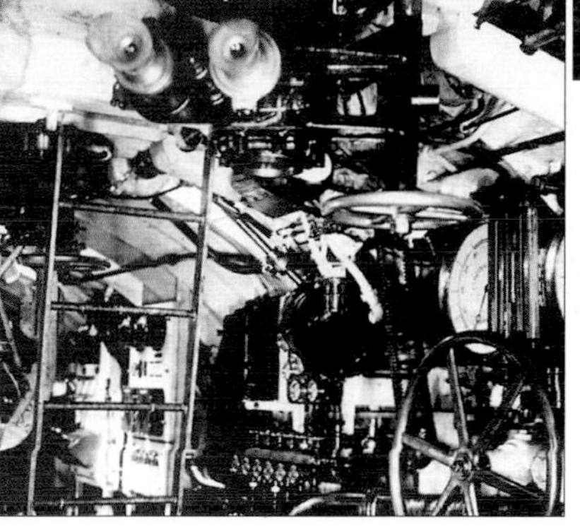

TOP AND ABOVE: **Commissioned in 1912, *Skipjack* (*E1*) had the distinction of being fitted with the first Sperry gyro compass in submarines, for which she became a pioneer underwater test ship.**

# E and F classes

These boats represented a substantial step forward in looks, style and design over their predecessors and were also the carriers of a number of US innovations. *E1* (laid down under the name of *Skipjack*) was commissioned in February 1912 under Lt Chester Nimitz, who also took command of the US Atlantic Submarine Flotilla before coming ashore in 1913 to continue his outstanding career in the big ships. *E1* was used to pioneer a number of major innovations, including the first ballistic gyro compass, invented by Elmer A. Sperry.

*E1* was also used to trial submerged radio transmission and other major experiments initially under Nimitz. During his time with the boat, he was awarded a Silver Medal for his heroic action in saving W.J. Walsh, fireman second class, from drowning. A strong tide was running and Walsh was rapidly being swept away from his ship. Nimitz dived into the water and kept Walsh afloat until both were picked up by a small boat.

In December 1917, *E1* left Newport for the Azores to begin patrols against U-boats, and at the end of the war, *E1* resumed her role in experimental work, conducting trials with undersea listening equipment.

The F-class boats were launched around the same time as the E class. The first, *F1*, was commissioned in June 1912 and was assigned to the 1st Submarine Group, Pacific Torpedo Flotilla, operating along the US west coast between San Diego and San Pedro until 1915, when the Flotilla moved to Honolulu for development operations in the Hawaiian Islands. Later, she returned to San Pedro to begin surface and submerged trials in the development of submarine tactics. On December 17, 1917, while manoeuvring in exercises at sea, *F1* and *F3* collided, the former sinking in 10 seconds, her port side torn forward of the engine room. Nineteen of her men were lost, while the others were rescued by the submarines with whom they were operating. Two E-class and four F-class submarines were built.

### *E1*

**Displacement:** 282 tonnes/277.5 tons (surfaced); 336 tonnes/330.7 tons (submerged)
**Length:** 41.24m/135ft 4in
**Beam:** 4.5m/14ft 9in
**Armament:** 4 x 457mm/18in tubes; 4 x torpedoes
**Propulsion:** 522kW/700hp diesel; 447.4kW/600hp electric
**Speed:** 13.5 knots (surfaced); 11.5 knots (dived)
**Complement:** 20 men

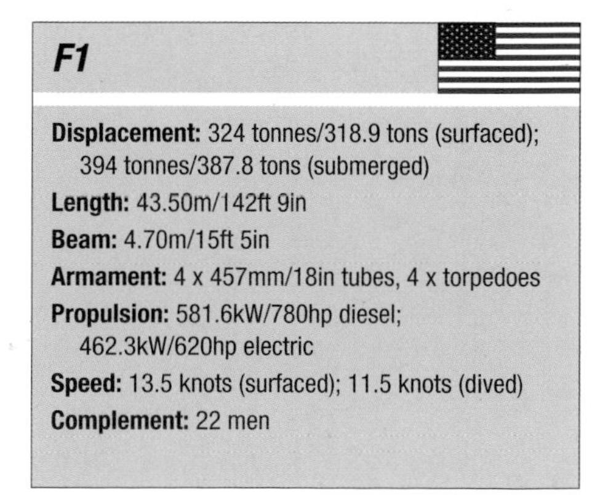

### *F1*

**Displacement:** 324 tonnes/318.9 tons (surfaced); 394 tonnes/387.8 tons (submerged)
**Length:** 43.50m/142ft 9in
**Beam:** 4.70m/15ft 5in
**Armament:** 4 x 457mm/18in tubes, 4 x torpedoes
**Propulsion:** 581.6kW/780hp diesel; 462.3kW/620hp electric
**Speed:** 13.5 knots (surfaced); 11.5 knots (dived)
**Complement:** 22 men

LEFT: **Simon Lake's *Seal (F1)* saw major advances in submarine technology in many ways, with innovations to improve comfort and safety.** ABOVE: **The launching of *Turbot* (*G3*) on December 27, 1913.**

# G class

*G1* has the distinction of being the first submarine designed for the US Navy by Simon Lake, the brilliant engineer from Pleasantville, New Jersey, whose submarine *Argonaut* had lost out to John Holland in the competition to build America's first boat. Rejected by his own country, he was poached by the Russians, then at war with Japan. He went on to build boats for several other nations. The American Navy chiefs thereafter shut him out of the competition for contracts until 1908 when, realizing the error of their ways, they finally approached him to build USS *Seal*, the first boat in G class.

It proved to be one of the most important developments in American naval history. Given virtually a free hand, Lake produced a revolutionary design for *Seal* (*G1*) in that it was the first-ever even keel submarine to be built by the United States Navy, and it was particularly noted for its sleek lines. However, it was only the beginning. Lake went on to initiate the introduction of many innovations in a long career producing boats for the United States, and his designs would influence submarine development well into the atomic age. It was only a pity that his enormous contribution was not formally recognized until after his death, in 1945.

*G1* joined the Atlantic Torpedo Flotilla and made a record dive of 78m/256ft in Long Island Sound. In October 1915, she began a new career as a submarine designated for experimental tests and instructional purposes. The G class thus became established in the important role of school ships for the newly established Submarine Base and Submarine School at New London. This was much needed as the submarine service expanded to meet the demands of World War I.

*G1* was also used to test new detector devices for the US Navy Experiment Board, off Provincetown, and played a similar role in the development and use of sound detection. Notably, she spent two four-day periscope and listening patrols against U-boats thought to be operating in the vicinity of Nantucket.

After the war, she resumed her role as a schooling boat for student submariners of the Listener and Hydrophone School, at New London, until January 13, 1920, when *G1* was towed to the Philadelphia Navy Yard. It was decommissioned on March 6, 1920.

She ended her days as a target for depth charge experiments by the Bureau of Ordnance before finally sinking on June 21, 1921 following a bombardment of eight attacks using experimental bombs.

LEFT: **A starboard side view of *Turbot* (*G3*), looking aft at the Lake Torpedo Boat Company shipyard, Bridgeport, Connecticut, 1915. Four G class were built.**

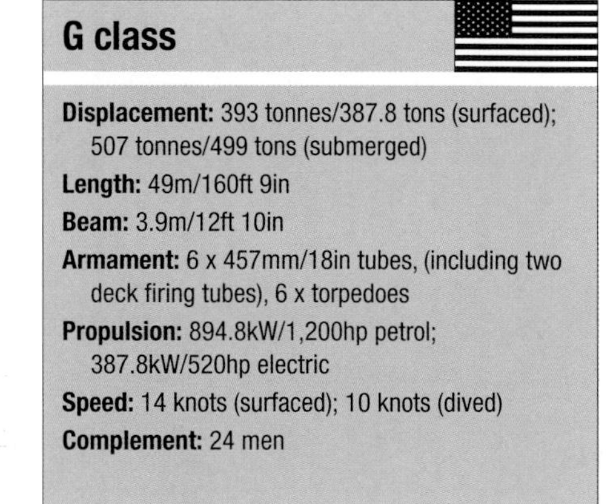

## G class

**Displacement:** 393 tonnes/387.8 tons (surfaced); 507 tonnes/499 tons (submerged)
**Length:** 49m/160ft 9in
**Beam:** 3.9m/12ft 10in
**Armament:** 6 x 457mm/18in tubes, (including two deck firing tubes), 6 x torpedoes
**Propulsion:** 894.8kW/1,200hp petrol; 387.8kW/520hp electric
**Speed:** 14 knots (surfaced); 10 knots (dived)
**Complement:** 24 men

# H class

*H1*, commissioned as *Seawolf*, was attached to the 2nd Torpedo Flotilla, Pacific Fleet, principally used for patrols along the west coast of America, as were other members of the class, operating out of San Pedro. When America entered World War I, the H boats were shifted to the east coast. *H1* completed the journey to New London 22 days later via Acapulco, Balboa, Key West, Charleston and Philadelphia. She remained on the east coast, patrolling Long Island Sound for the duration, often with officer students from the submarine school on board. *H1* and *H2* (originally named *Nautilus*) returned to San Pedro in 1920 via the Panama Canal but on March 12, *H1* ran aground on a sandbank off Santa Margarita Island, California. *H2* stood by and sent rescue and search parties for survivors, helping to save all but four of her sister ship's crew. Those killed included the commanding officer, Lt Commander James R. Webb. A salvage ship dragged her clear only to see her sink in 18.3m/60ft of water.

ABOVE: **USS *Garfish* (*H3*) in dry dock at the Mare Island Navy Yard, California, in 1915, to repair damage following a recent grounding, evident on the bilge keel.**
LEFT: ***H2*, the first US submarine to bear the name *Nautilus*.**

A similar misfortune hit *H3*. While engaged in patrolling the Californian coast, near Eureka, she ran aground in heavy fog on December 16, 1916. The crew was rescued by Coast Guard breeches buoy, but the cruiser *Milwaukee* was also stranded trying to pull the submarine off the beach, and the Navy was forced to call in a commercial salvage firm. Their job was made all the more difficult because *H3* was marooned on a high sandbank surrounded by quicksand, and at low tide she was 22.9m/75ft from the water, but at high tide the ocean reached almost 76.2m/250ft beyond her. Eventually, after a month of trying, she was pulled on to giant log rollers and taken overland to the sea.

Six other H-class submarines originally built by the Electric Boat Company for the Imperial Russian Government also came into the US inventory under the H4 class title. Their shipment was held up pending the outcome of the Russian Revolution, and the boats were stored in knockdown condition at Vancouver, B.C. All six were eventually purchased by the Navy on May 20, 1918, and assembled at Puget Sound Navy Yard. The first went into service in November 1918, and were engaged in extensive battle and training exercises out of San Pedro, varying this routine with patrols off Santa Catalina Island, but by then, of course, the Armistice had been signed. They were decommissioned in the early 1920s and eventually sold for scrap.

LEFT: **USS *Sea Wolf* (*H1*) was the prototype for a class that attracted considerable orders from overseas, yet only three were built for the US Navy. Most were destined for Britain in the latter stages of World War I.**

### H class

**Displacement:** 352 tonnes/346.4 tons (surfaced); 459 tonnes/451.8 tons (submerged)
**Length:** 45.82m/150ft 4in
**Beam:** 4.84m/15ft 11in
**Armament:** 4 x 457mm/18in tubes, 8 x torpedoes
**Propulsion:** kW/950hp diesel; kW/600hp electric
**Speed:** 14 knots (surfaced); 10.5 knots (dived)
**Complement:** 25.men

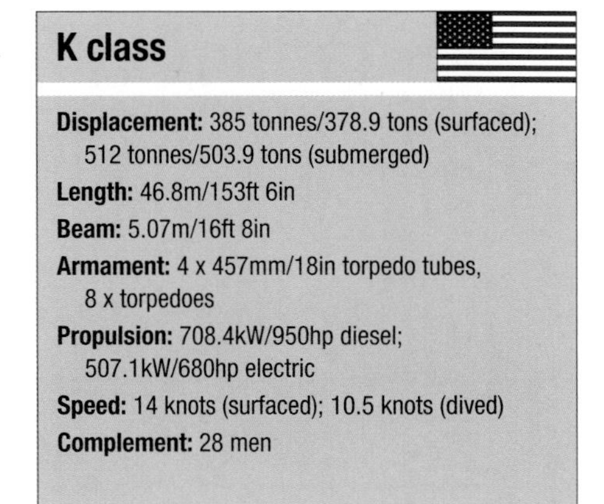

TOP: **The US Navy dropped the practice of naming their boats in the K class, retaining the class number and boat number, in this case *K8* and *SS39*, here dry-docked at Honolulu.** ABOVE: ***K5* underway on a 1919 cruise down the Mississippi River.**

# K class

K1, commissioned as *Haddock*, and her sister ships continued the emerging theme in US submarine design: to increase and improve space, conditions and safety inside the boats without a massively increased overall displacement or length. This class was no exception, increasing the complement to 28 men with the addition of little more than 30 tonnes/29.5 tons surface displacement. The K boats were deployed on both the east and west coasts of America, mostly in almost continual trials in underwater manoeuvres, diving and torpedo-firing practice. The techniques learned from these experiments were soon to prove useful when U-boats began their attacks on Allied shipping bound for Europe.

Five of the K boats were among the first US submarines assigned to duties in World War I and saw service off the Azores, conducting patrol cruises and protecting shipping from surface attack. Some boats were modified for these operations, including the introduction of passive sound gear forward, a permanent chariot bridge and housing periscopes. These vital patrols continued almost to Armistice Day, by which time they returned to North America to resume coastal operations. They were then assigned to naval exercises and trials of new technology that contributed much to the improvements that lay ahead.

The K-class boats were especially engaged in experimental work, notably in the areas of listening devices, storage batteries and torpedoes, and the *Dictionary of American Naval Fighting Ships* states that this work "greatly accelerated" the technology learned from these experiments. The boats were gradually decommissioned in the early 1920s. Eight K-class boats were built.

LEFT: ***K2* off Pensacola, Florida, in April, 1916, in a new "zebra" camouflage.**

## K class

**Displacement:** 385 tonnes/378.9 tons (surfaced); 512 tonnes/503.9 tons (submerged)
**Length:** 46.8m/153ft 6in
**Beam:** 5.07m/16ft 8in
**Armament:** 4 x 457mm/18in torpedo tubes, 8 x torpedoes
**Propulsion:** 708.4kW/950hp diesel; 507.1kW/680hp electric
**Speed:** 14 knots (surfaced); 10.5 knots (dived)
**Complement:** 28 men

# L class

ABOVE: **A profile of the L class, the first of which began operations in the US in 1916. It became the largest class of American submarines to date, with 11 being built for service in World War I.**

*L1* was christened as *SS40* when the new numbering classification replaced the tradition of naming the boats came into effect with the last three of the K boats. This was to last until names, as well as numbers, were introduced with the Barracuda class in the early 1920s. Again, displacement and length were further extended, and wartime modifications to the class included the installation of a retractable mount for the 76mm/3in gun, and sound gear as well as a permanent sheltered bridge. The retractable gun mount was an idea inspired by a version seen on board a pre-war U-boat.

The L boats were also the first to use independent torpedo tube shutters, which replaced a single-rotating bow cap. A further interesting aspect of the Ls was the difference between those designed and built by the Electric Boat Company and those developed by Simon Lake. The latter's boats were more compartmentalized in terms of the mechanics of the boat than the Electric Boat Company vessels, providing separate engine and motor rooms aft, and concentrating pumps and other auxiliary tools into the large space beneath the control room.

For their war service, L boats crossed the Atlantic after brief service around the Azores and operated from Bantry Bay, Ireland, where they began patrols in support of Allied shipping and hunting U-boats. In July 1918, a large explosion rocked *L2* while patrolling in the Irish Sea. A periscope was sighted and the *L2* submerged and tried to ram the submarine, but the U-boat had superior underwater speed. Later, it was suspected that a U-boat had fired on the *L2*, but another U-boat, the *U65*, was in the way and was badly damaged and sank. Some time later, when the *L2* was dry-docked, her hull plating was noted to be heavily dented from the proximity of the explosion. The *U65* never returned to port.

Meanwhile, in the US, one of the Lake boats, *L8*, was seconded to a top secret mission initiated in 1918 to lure U-boats into a trap. The boat joined forces with the schooner USS *Charles Whittemore*, which was to act as a decoy in the hope of prompting an attack from a pair of enemy submarines menacing the Atlantic coast. The *Whittemore* towed the submerged *L8* to avoid the submarine's detection, while at the same time acting as mother ship, carrying supplies, fuel and torpedoes. In the event, the war ended while the pair was on patrol without firing a shot. Eleven L boats were built and they were decommissioned progressively in the mid-1920s.

ABOVE: **Submarine bridges of the day were small to limit underwater drag. *L3* had an enlarged chariot bridge, in contrast to the more streamlined *L9*.**
LEFT: ***L10* arrived in the British Isles in January 1918 with other submarines of the Atlantic Division to take part in operations hampering U-boat activity.**

## L class

**Displacement:** 443 tonnes/436 tons (surfaced); 539 tonnes/530.5 tons (submerged)
**Length:** 51m/167ft 4in
**Beam:** 5.3m/17ft 5in
**Armament:** 4 x 457mm/18in tubes, 8 x torpedoes, 1 x 76mm/3in (23 calibres) deck gun
**Propulsion:** 671.1kW/900hp diesel; 507.1kW/680hp electric
**Speed:** 14 knots (surfaced); 10.5 knots (dived)
**Complement:** 28 men

# M class

*M1* was something of an experimental boat, and as only one of the class was built, it was generally assumed that there were flaws that prevented further orders. Doubtless there were flaws, but she was rightly in a class of her own. *M1* was the longest boat built to that point in time (1918) and was the Electric Boat Company's first US double-hull submarine. The company had already designed larger double-hull boats for Russia, which were a match for the cruiser-style boats coming out of Germany. In her trials, she was promoted as being the embodiment of all the newest technology in submarine construction and design. Her battery was to have solved some of the past difficulties in submarine battery design and operation.

With a stern described as being honed to a vertical chisel shape, much like contemporary cruiser sterns, she was an impressive sight. Inside, there was decent – if somewhat crowded – accommodation for the crew. Indeed, a Royal Navy observer, Stanley Goodall, later knighted as head of British naval construction, noted that facilities were very presentable, with bunks stacked three high that were light and easily stowed. An additional luxury was that the boat was heated and had an ice tank for food storage.

Even so, *M1* had a short life of barely four years' active service post-war, before being decommissioned and scrapped.

LEFT: ***M1* was the Electric Boat Company's first US double-hull submarine, although the company had already designed a larger double-hull boat for Russia. This submarine also had a number of internal innovations, including an efficient heating system and an ice box/refrigerator.**

### M class

**Displacement:** 480 tonnes/472.4 tons (surfaced); 665 tonnes/654.5 tons (submerged)
**Length:** 58.75m/192ft 9in
**Beam:** 4.49m/14ft 9in
**Armament:** 4 x 457mm/18in tubes, 8 x torpedoes, 1 x 76mm/3in (23 calibres) deck gun
**Propulsion:** 626.4kW/840hp diesel; 507.1kW/680hp electric
**Speed:** 14 knots (surfaced); 10.5 knots (dived)
**Complement:** 28 men

LEFT: **After war service patrolling the New England coast of America, *N2* served as a training ship for the Submarine School and in 1921 became a test platform for experimental Navy weapons, such as a radio controlled torpedo.**

### N class

**Displacement:** 342 tonnes/336.6 tons (surfaced); 407 tonnes/400.6 tons (submerged)
**Length:** 44.85m/147ft 2in
**Beam:** 4.8m/15ft 9in
**Armament:** 4 x 457mm/18in tubes, 8 x torpedoes
**Propulsion:** 357.9kW/480hp diesel; 208.8kW/280hp electric
**Speed:** 13 knots (surfaced); 11 knots (dived)
**Complement:** 25 men

# N class

These boats were at the other end of the spectrum compared to *M1*, being relatively small and intended largely for harbour defence and even inland waterway patrols. *N3* became one of the first submarines to navigate the St Lawrence River and the Great Lakes and visited Halifax, Quebec, Montreal, and Port Dalhousie before arriving at Toledo on June 25, 1921.

At most stopping points, she went on public display and took visitors aboard. Rushed into being towards the end of World War I to meet production targets, none saw any challenging duties in the few remaining months of the conflict.

Thereafter the N-class boats were used extensively as training platforms, given the absence of any serious threats in the intended role. Like the M class, they too had a short life. Seven of the class were built, but virtually all were decommissioned by the mid-1920s.

# O class

After the small harbour protectors came the O-class boats, 500 tonnes/492 tons or more in surface displacement. These were laid down before America's involvement in World War I in sufficient numbers to provide a failsafe position should the worst happen – as, indeed, it did. It was in many ways rather ironic that the O boats began rolling off the production lines, along with the even more advanced R class, providing some satisfaction for the public and the media clamouring for new ships and submarines. In the event, the boats of this class saw little or no war service – at least not in the way they were intended. On November 2, 1918, a number of the Os were included in a 20-boat contingent that set out for European waters to take on the U-boats, but the Armistice was signed before the ships reached the Azores, and they returned to the United States without firing a single torpedo in anger.

It was a remarkable fact that, while many boats were disposed of during the 1920s and early 30s for scrapping – some in connection with the international agreements to limit submarine construction — eight of the O class were mothballed. These were brought back into service following the Japanese attack on Pearl Harbor in 1941, to become the oldest boats in the US submarine fleet to participate in World War II. Most survived the war, remaining in service throughout, fulfilling operational and training purposes, with two still in business in 1946.

ABOVE: **The O-class side view.** RIGHT: **After 12 years' service in her duties for the US Navy, *O12* was converted for use on the Sir Hubert Wilkins Arctic Expedition of geophysical investigation, for which she was given the temporary name of *Nautilus*. After being returned to the Navy, she ended up in a Norwegian fjord, where she sank in 1931.**

There was one casualty among them, occurring in the months prior to America entering World War II. Recommissioned in April 1941, *O9* left New London with others of her class on June 19 for trials off the Isles of Shoals, 24km/15 miles off Portsmouth, New Hampshire. The following day, after sister ships *O6* and *O10* had successful completed their trials, *O9* hit trouble during deep submergence tests and slipped well below her depth limit of 60.9m/200ft. She was crushed by the pressure of the water and sank. Search operations were launched immediately by the sister ships and one of the newer submarines, *Triton*, along with the submarine rescue vessel, *Falcon*, but they found no trace.

After two days, divers eventually located her in 134m/440ft of water in the same area where *Squalus* had been lost in 1939. The depth meant they could only stay down for a brief time and salvage operations were considered too risky. All further efforts to raise her were cancelled and the crew of 33 was declared lost.

ABOVE: **The starboard diesel engine of the O class, which was upgraded as the building of the O boats progressed to the last of that configuration, *O16*, which had Busch Sulzer Brothers diesels to provide 745.7kW/1,000hp driving a single propeller.**

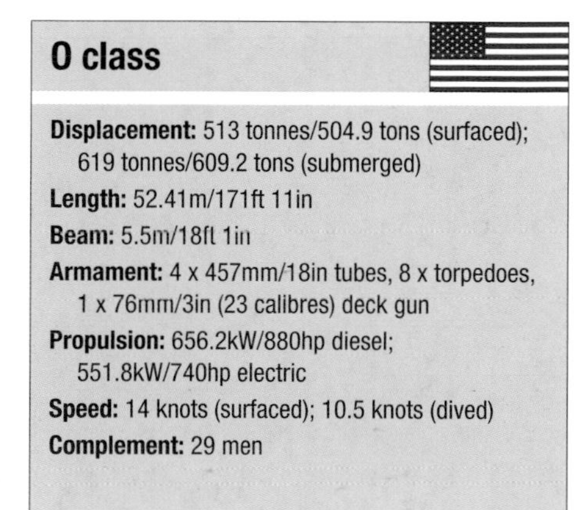

## O class

**Displacement:** 513 tonnes/504.9 tons (surfaced); 619 tonnes/609.2 tons (submerged)
**Length:** 52.41m/171ft 11in
**Beam:** 5.5m/18ft 1in
**Armament:** 4 x 457mm/18in tubes, 8 x torpedoes, 1 x 76mm/3in (23 calibres) deck gun
**Propulsion:** 656.2kW/880hp diesel; 551.8kW/740hp electric
**Speed:** 14 knots (surfaced); 10.5 knots (dived)
**Complement:** 29 men

# R class

The R class had a very similar story to that of the O boats, whose relatively short working life in the 1920s was also reactivated for World War II. This class also came into being as World War I was ending, with no fewer than 27 having been laid down in 1917 for commissioning by the end of 1918. At that time, plans for the much larger new S class were also on the drawing board. By then the emergency was over, and with such a surfeit of submarines available to the United States as the much larger S class also came into being, many of the R boats were laid up by the mid-1920s, and all had been decommissioned by 1931 in the wake of international agreements on undersea warfare.

The R class was built by three different submarine yards to achieve prompt delivery in case the war dragged on, and there were some differences in the final build, in terms of compartments and even overall length and weight. The first years of service for the class leader *R1* and many of her sister boats were spent on patrol, training crews and developing submarine tactics before being laid up. She was brought back to life in 1940 and, after a refit, was transferred to Squadron 7 to hunt U-boats on a patrol line 402km/ 250 miles north-east of Bermuda, a task which continued through World War II. From 1944, she and several sister ships were tasked exclusively with anti-submarine work.

The rest of the R-class boats followed much the same pattern, and suffered only two losses. *R12*, which was recommissioned in October 1940, spent much of her time patrolling the region around the entrance to the Panama Canal, operating primarily from Guantanamo Bay. However, on June 12, 1943 while conducting torpedo practice with Brazilian observers on board, the forward battery compartment flooded as she prepared to dive. She sank in less than half a minute taking the 42 officers, crew and visitors with her. None escaped.

The demise of *R1*, on the other hand, was down to a case of mistaken identity. She was among the boats transferred to the Royal Navy under a lend-lease programme in 1942 and reclassified as *P514*. Unfortunately, she was rammed by HMCS *Georgian* on June 21, 1942 in the western Atlantic. The Canadians thought she was a U-boat. She sank in minutes with all hands.

ABOVE LEFT AND ABOVE: **Six R boats and two S class boats nestled together off New York City, in May 1920, alongside a submarine tender. All of the R boats have gun platforms, but guns are fitted only on *R10* and *R3*.**

ABOVE: ***R2*, like many of her class of 27 boats, took a long layoff from duty for much of the 1930s, having been decommissioned only to be brought back into service in the period immediately prior to World War II.**

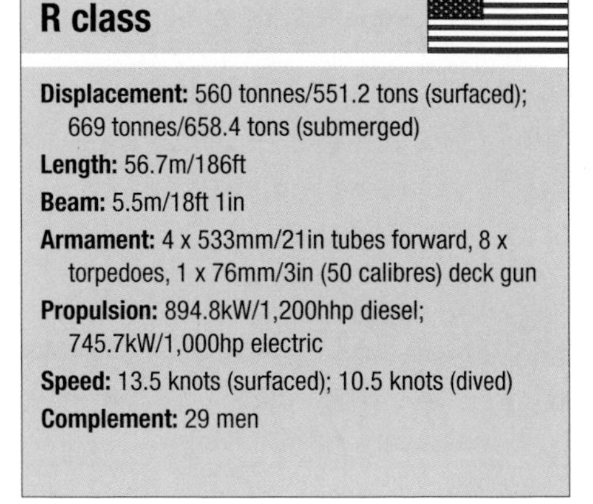

**R class**

**Displacement:** 560 tonnes/551.2 tons (surfaced); 669 tonnes/658.4 tons (submerged)
**Length:** 56.7m/186ft
**Beam:** 5.5m/18ft 1in
**Armament:** 4 x 533mm/21in tubes forward, 8 x torpedoes, 1 x 76mm/3in (50 calibres) deck gun
**Propulsion:** 894.8kW/1,200hhp diesel; 745.7kW/1,000hp electric
**Speed:** 13.5 knots (surfaced); 10.5 knots (dived)
**Complement:** 29 men

LEFT: ***S40*** **was typical of many of her class in that they were excellent submarines built for work in the 1930s but not quite robust enough for what was to follow in the 1940s. Even so, collectively they had a fine record,** ***S40*** **completing nine war patrols, operating out of the East Indies and Australia.**
BELOW: **An impressive "sharp end" view of** ***S4*** **exiting the repair bay.**

# S class

S boats were built in three types in the 1920s under the same general specifications but with varying designs. However, as an overall mass-production class, it was the largest class ever undertaken by the US Navy, in every respect. Fifty-one boats of this class were built and, becoming America's most successful submarine to date, they were operational in each of the following three decades. The first, *S1* – known as the Holland type – saw the culmination of the Electric Boat Company's single-hull design for the US Navy. *S2* was known as the Lake-type and *S3* as the Government-type. All three versions undertook experimental duties of varying kinds in the decade of their manufacture, which greatly contributed to future US submarine developments. One of particular note fell to *S1* in that she became the experimental platform for the US Navy's first submarine aircraft carrier. This would also be tried by other countries, notably Britain, Italy and Japan, but only the latter continued the experiments into wartime operational use of any magnitude.

*S1* was given a cylindrical capsule mounted to the rear of the conning tower to house a collapsible Martin MS-1 seaplane, first built of wood and fabric and later in two metal types. On the surface, the aircraft would be rolled out, assembled and launched from the deck by partially submerging the submarine. The first trials that took the operations through all phases, from loading the disassembled aircraft into its pod through to reconstruction at sea, launching and eventual retrieval were conducted in July 1926 on the Thames River at New London.

*S1* then went on to become a divisional flagship operating in the Panama Canal Zone and later took up duties based at Pearl Harbor and in other home ports. By then, a fairly active peacetime training cycle had been established across the board so all boats took up rotational postings around all areas of operation. These preparations did not stop the Japanese attack on the Pearl Harbor base, but did however prepare the S boats for a staunch and versatile involvement in World War II, from defensive and offensive positions.

Six of the S boats became casualties: *S5* sank after an accidental intake of water soon after being commissioned, but miraculously all aboard escaped through a hole cut in the hull. *S26* sank after a collision in the Gulf of Panama in December 1941, with the loss of 46 men; three survived. On September 25, 1925, *S51* was rammed and sunk off Block Island, New York, by a merchant steamer, with only three survivors. *S28* was lost at sea with all hands in July 1944. *S39* became a victim of accidental grounding in 1942 after heroic action against the Japanese, but the entire crew was evacuated to safety. *S44* was the victim of heavy enemy bombardment while stranded through mechanical problems on the surface in October 1943. Only two of the crew survived. They spent the rest of the war in a Japanese slave labour camp.

LEFT: **The S-class profile, which remained virtually the same for all of its 51 boats.**

## S class

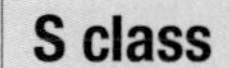

**Displacement:** 840 tonnes/826.7 tons (surfaced); 1,045 tonnes/1,028.5 tons (submerged)
**Length:** 66.75m/219ft
**Beam:** 6.35m/20ft 10in
**Armament:** 4 x 533mm/21in tubes, 12 x torpedoes, 1 x 102mm/4in (50 calibres) deck gun
**Propulsion:** 1,342.2kW/1,800hp diesel; 1,118.6kW/1500hp electric
**Speed:** 14.5 knots (surfaced); 11 knots (dived)
**Complement:** 38 men

LEFT: **By now, the US Navy had reverted to names rather than numbers, and *Skipjack* came in the Salmon class, a strong class that showed influences from the recent trials with larger submarines such as the V boats *Barracuda*, *Argonaut* and *Narwhal*.** ABOVE: **The *Skipjack* badge.**

# Salmon class

In the 1930s, the classification of new submarines in the US Navy reverted to the name of the lead boat, in this case *Salmon*, commissioned in 1938. This class was one of the new generation of huge submarines that had begun rolling out of the US shipyards in the wake of experimental fleet and cruiser submarines of the late 1920s, such as *Argonaut*, *Narwhal* and *Nautilus*, originally commissioned under the V classification. Indeed, from the mid-1930s and beyond, a massive production schedule was imposed upon the shipbuilders as the US government embarked upon a programme of submarine construction matched at that time only by the Japanese. *Salmon*, laid down in April 1936 by the Electric Boat Company at Groton, Connecticut, was commissioned 23 months later, and was joined over the coming months by five others in her class, names which became famous in the annals of US submarine history – *Seal*, *Skipjack*, *Snapper*, *Stingray* and *Sturgeon*.

After coastal trials from Nova Scotia to the West Indies, *Salmon* became flagship of Submarine Division 15, Squadron 6 of the Submarine Force, US Fleet, at Portsmouth, operating along the Atlantic coast for a year. She then transferred to the west coast to bolster the submarine presence prior to transferring to the Asiatic station in Manila, along with submarine tender *Holland* and three other boats, to strengthen defences around the Philippines in response to growing tension with Japan. *Salmon* had been leading defensive patrols from Manila when the Japanese attacked Pearl Harbor on December 7, 1941, and, like US submarines everywhere, was placed on an immediate offensive alert and a few days later fired her first shots in anger, sending a spread of torpedoes against Japanese destroyers to scatter their attack.

As the opposition became fierce in the region, the submarine base was moved out of Manila to Exmouth Gulf, Australia, from where sorties into the Java Sea were launched. In March, *Salmon* moved to the US submarine base at Fremantle, Australia, from where patrols were being organized and established along the south coast of Java to intercept Japanese shipping. On May 3, she torpedoed and sank the 11,625-tonne/11,441-ton repair ship *Asahi*, and on the 28th, sank the 4,452-tonne/4,382-ton passenger-cargo vessel, *Ganges Maru*. These operations were merely the beginning of a busy wartime career operating from both Australia and Peal Harbour, for which she was awarded nine battle stars for Asiatic and Pacific operations.

ABOVE: **A bow view of USS *Snapper* in dry dock in 1939. It was a fortuitous refit for what lay ahead, which in this boat's case amounted to 11 busy war patrols, for which she was awarded six battle stars.**

## Salmon class

**Displacement:** 1,426 tonnes/1,403.5 tons (surfaced); 2,163 tonnes/2,128.8 tons (submerged)
**Length:** 93.8m/307ft 9in
**Beam:** 7.9m/25ft 11in
**Armament:** 8 x 533mm/21in tubes, 24 x torpedoes, 1 x 76mm/3in (50 calibres) deck gun, 2 x 12.7mm/0.50 calibre machine-guns, 2 x 7.62mm/0.30 calibre machine-guns.
**Propulsion:** 4,101.3kW/5,500hp diesel; 4 x electric motors developing 2,460.8kW/3,300hp
**Speed:** 21 knots (surfaced); 9 knots (dived)
**Complement:** 55 men

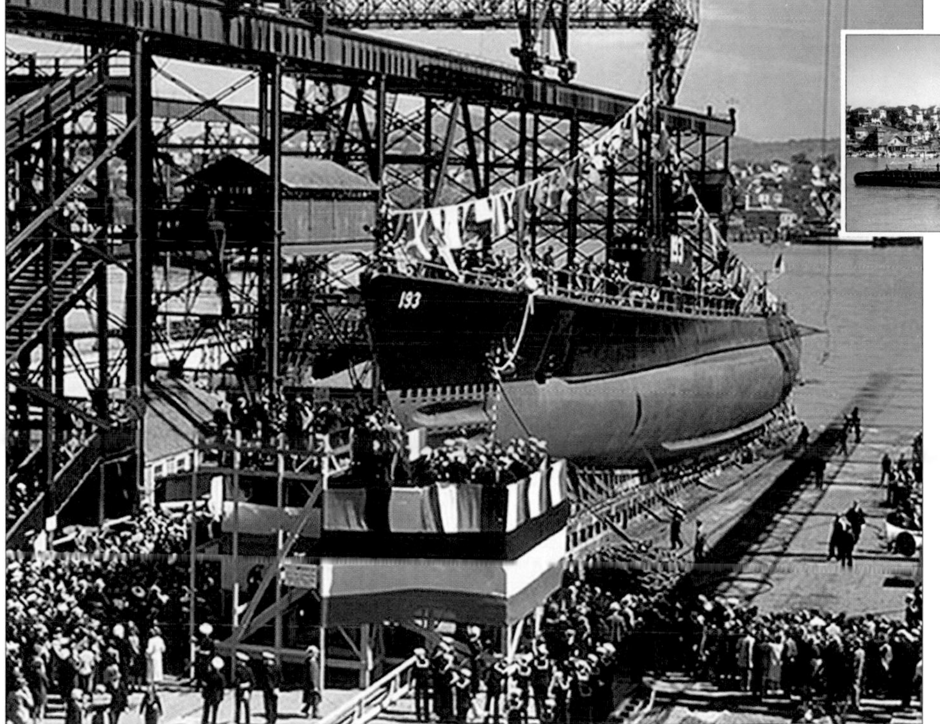

LEFT: **The launch of *Spearfish*, third of the ten-boat Sargo class, all of which were in for a busy war. Apart from aggressive patrolling, *Spearfish* was also engaged on dangerous reconnaissance work, winning ten battle stars.** ABOVE: **Sister ship *Swordfish*, launched in 1939.** BELOW: **The war is over in this photograph, which includes class leader *Sargo* with *Spearfish* and *Saury*. The mothballing process is underway with the preservative cocoons around the deck guns that would never again be fired in anger.**

# Sargo class

Ten boats from this class were commissioned progressively from February 1939 to become the first of the principal classes in the US Navy's submarine fleet amassed before and during World War II – a force which amounted to more than 400 boats in addition to another 100 from the classes listed previously. *Sargo* went straight to business with a war patrol off French Indochina and to the Dutch East Indies, delivering eight separate attacks on enemy shipping. However, her Mark 14 torpedoes malfunctioned, and her targets escaped. Later that year (1942), she temporarily took on the role of transport, offloading her remaining torpedoes and 76mm/3in ammunition to stow a million rounds of 7.62mm/0.30 calibre ammunition desperately needed by Allied forces in the Philippines, and thereafter took up station at Fremantle.

The welcome she received was somewhat unfortunate. As the first Sargo class in the region, she was mistaken for a Japanese vessel and attacked by an Allied aircraft. Fortunately, the damage was not excessive and she went on to participate fully in the war effort with ten patrols, receiving the Philippine Presidential Unit Citation for her work in that area.

However, another blue-on-blue incident had a devastating effect on another in the Sargo class. *Sculpin* was scheduled to lead a co-ordinated group attack on enemy shipping located near the Eniwetok islands in the Pacific. During the raid, *Sculpin*, already at her depth limit, was blasted by 18 depth charges and with her pressure hull distorted, the boat leaking and diving plane gear damaged, Captain Fred Cromwell had no option but to surface. As her executive officers came on to the bridge, they were killed by shellfire, some of which went through the main induction. Cromwell could do nothing more, and ordered his submarine to be scuttled and all hands to abandon ship.

About 12 men "rode the ship down" according to an account in the *Dictionary of American Naval Fighting Ships*. Captain Fred Cromwell stayed aboard, determined to go down with his submarine, because he alone possessed secret intelligence of US Fleet movements and specific attack plans. He feared that he would be forced to reveal this information under Japanese torture or use of drugs, an action for which he was posthumously awarded the Congressional Medal of Honour.

In all, 42 men were taken prisoner by the Japanese destroyer. The group was split up, 21 in the carrier *Chuyo* and 20 in another ship, bound for captivity. Now, an ironic twist of fate took hold: en route to Japan, *Chuyo* was torpedoed and sunk by another *Sargo*-class boat, *Sailfish*, on December 31, 1943.

## Sargo class

**Displacement:** 1,427 tonnes/1,1404.5 tons (surfaced); 2,312 tonnes/2,275.5 tons (submerged)
**Length:** 94.5m/310ft
**Beam:** 8.25m/27ft
**Armament:** 8 x 533mm/21in tubes, 24 x torpedoes, 1 x 76mm/3in (50 calibres) deck gun, 2 x 12.7mm/0.50 calibre machine-guns, 2 x 7.72mm/0.30 calibre machine-guns
**Propulsion:** 4,101.3kW/5,500hp diesel; 2,043.2kW/2,740hp electric
**Speed:** 20 knots (surfaced); 7.75 knots (dived)
**Complement:** 55 men

# Gato class

*Gato* had a baptism of fire as one of the boats commissioned in December 1941 soon after the Japanese attack on Pearl Harbor. She was the first of 74 boats in this class, one of three main types that became the workhorses of the American effort in World War II. Developed towards the end of the 1930s, and employing the round-the-clock capacity of every major submarine builder in the United States, the new submarines were coming off the production lines at an incredible rate, a state of affairs that was to continue well into 1943.

Given that the surface fleet was being similarly enhanced, this represented an outstanding effort by the shipbuilders, who were producing brilliant boats that incorporated numerous technological advances covering every aspect of submarine activity. These boats bristled with new and heavy deck armoury and carried a substantial stock of torpedoes to match their increasing endurance. The 1,500-tonne/1,476-ton (surfaced) Gato class could stay out on patrol for 74 days, and cruise 17,703km/11,000 miles surfaced at 10 knots, although the boats were capable of a top speed of 20 knots.

It was also a great credit to their designers, engineers and shipwrights that, unlike the many submarines produced up to that point in time, whose life expectancy was often quite limited, the Gato class led the way to longevity. Along with the other newer classes that followed, many boats were

ABOVE: **Larger still, in class and size, *Raton* is photographed here in post-war mode as a Radar Picket submarine, with a slightly different configuration, circa 1953–60.** LEFT: **The famous *Wahoo,* one of the US Submarine Force's most successful boats, whose exploits under Commander "Mush" Morton were legendary. During just six patrols she sank 27 ships, totalling 121,011 tonnes/ 119,100 tons, and damaged two more, making 25,300 tonnes/ 24,900 tons before her own luck ran out on October 11, 1943, when she was lost with all 79 hands.**

still in service in the 1970s, some of which were converted into the post-war classification of hunter-killers in the Guppy configuration.

Despite their increasing size, the versatility of submarines was also being extended to the full, especially in the areas of intelligence-gathering and combined operations with special forces. *Gato* herself, for example, did some fine work landing Australian commandos in Japanese-held territory while at the same time transporting evacuees to safety. In fact, the submarine survived 13 war patrols, her last a particularly dangerous mission, rescuing downed Army aviators after waves of air strikes off the eastern coast of Honshu. It was there that she received the signal "Cease Fire" on August 15, 1945, and then steamed into Tokyo Bay. She remained there for the signing of surrender documents on board USS *Missouri* on September 2, before setting off for home to receive the Presidential Unit Citation in recognition of daring exploits, and 13 battle stars for service in World War II.

In all, the *Gato*-class boats won more than 700 battle stars, although the cost was high, with the loss of 19 boats and more than 1,100 men to wartime action. The casualties included a particular, but not especially uncommon, misfortune that befell the last of the class, *Tullibee*, which occurred in 1943 while the boat was patrolling off the Philippines. She had fired a volley of torpedoes towards an enemy passenger ship and two destroyers, at least two of which were hit. Minutes later an explosion rocked the submarine and Gunner's Mate C.W. Kukyendall – on the bridge at the time – was knocked unconscious and thrown into the water. When he regained consciousness, the submarine was gone. One of her own torpedoes had apparently turned full circle and hit her. Kukyendall was the only survivor.

LEFT: **The patches for class leader USS *Dace* and USS *Gato*. Both boats survived the war with commendations for meritorious service.**

BELOW: **USS *Barb* had a busy after-life following her war service, first undergoing a Guppy conversion and later on being loaned to Italy in 1955, where she served under the name of *Enrico Tazzoli* until 1975.** BOTTOM: **A slightly cramped-looking control room in USS *Cero*, one of a large number of boats laid up in reserve after the war.**

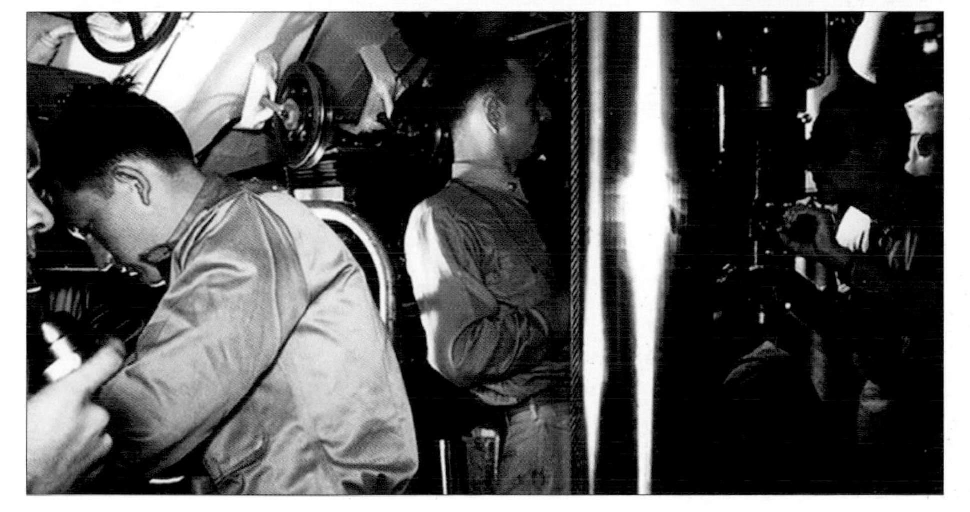

## Gato class

**Displacement:** 1,501 tonnes/1,477.3 tons (surfaced); 2,386 tonnes/2,348.3 tons (submerged)
**Length:** 95m/311ft 8in
**Beam:** 8.3m/27ft 3in
**Armament:** 10 x 533mm/21in tubes, 6 x forward, 4 x aft, 24 x torpedoes, 1 x 76mm/3in (50 calibres) deck gun, 2 x 12.7mm/0.50 calibre machine-guns, 2 x 7.63mm/0.30 calibre machine-guns
**Propulsion:** 4,026.8kW/5,400hp diesel; 2,043.2kW/2,740hp electric
**Speed:** 20.25 knots (surfaced); 8.75 knots (dived)
**Complement:** 60 men

LEFT: **More big boats and a big history in the Balao class, this one being *Pampanito*, noted for her role in the American submarine service's own wolf-pack operations in the South China Sea, although she had to temporarily retire for repairs after a severe depth charge attack.**

# Balao class

This was the largest class of submarines ever produced by the US Navy, in terms of numbers, with a total of 131 earmarked for production during World War II. Ten were subsequently cancelled towards the end of the production cycle, by which time the war's end was in sight. From first to last in production – *Balao* to *Tiru* – was a three-year span from the point in April 1942 when the class leader was laid down, to 1947 when the last was commissioned. *Balao*'s own war operations spanned a period from July 1943 until the end of August 1945. During this period, she completed ten war patrols and ended her war effort with a fine record, having sunk seven Japanese ships totalling 32,623 tonnes/32,108 tons, in addition to sinking 11,177 tonnes/ 11,000 tons of miscellaneous enemy small craft by gunfire.

As a whole, the Balao class, with the largest number of submarines afloat during the final 18 months of the war, undertook many successful patrols in all regions of the conflict. However, the cost was high. Nine boats and almost 600 men were lost to enemy action. Throughout the submarine stories, there were many examples of heroism and survival that provided dramatic true-life scenarios for numerous movies. One such event was recognized in 2004 with the erection of a plaque at the Damage Control Training Centre at Pearl Harbor, a story from November 1943 aboard *Balao*-class *Billfish* that led to extensive research in damage limitation. The submarine became riddled with leaks after being inundated with depth charges during an attack by the Japanese that incapacitated the boat's commander and senior officers.

As the attack continued, Lt Charles Rush, then 23, took charge and risked diving the boat to 51.8m/170ft below test depth for 12 hours. Meanwhile Chief Electrician's Mate John D. Rendernick gathered a team for emergency repairs to reduce the heavy flooding through the stern tubes by pumping grease into the worst leaking tube. He then used six men and a hydraulic jack to reposition the port main motor, which had been knocked off its foundation. When the increasing heat reached the limit of the crew's tolerance, Rendernick told them to wrap wet towels around their heads. These actions and Rush's coolheaded command in the face of continuing explosions around the boat, saved it from destruction – and the lives of the entire crew. Crucially, he also turned his boat to regain a previous position to sit under an oil slick, thus fooling the Japanese destroyers above into thinking the boat had been destroyed.

After the war, many of the surviving Balao class underwent substantial remodelling, both internally and in the superstructure to keep them operational for years to come. *Balao* was among a number of boats sold to Turkey under a Security Assistance Program in 1970, and under the new name of *Burak Reis*, she remained on the Turkish naval register until 1996.

LEFT: **USS *Blackfin*, one of nine Balao submarines converted to the Guppy 1A configuration between 1947 and 1951. This was a less drastic remodelling than the costly Guppy II programme, an interim measure in the developing Cold War providing boats with streamlined sail and superstructure, new masts, snorkel and equipment, and greater underwater propulsion.**

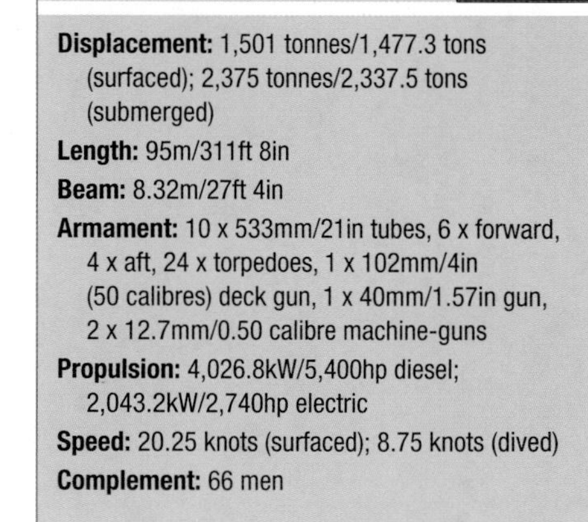

## Balao class

**Displacement:** 1,501 tonnes/1,477.3 tons (surfaced); 2,375 tonnes/2,337.5 tons (submerged)
**Length:** 95m/311ft 8in
**Beam:** 8.32m/27ft 4in
**Armament:** 10 x 533mm/21in tubes, 6 x forward, 4 x aft, 24 x torpedoes, 1 x 102mm/4in (50 calibres) deck gun, 1 x 40mm/1.57in gun, 2 x 12.7mm/0.50 calibre machine-guns
**Propulsion:** 4,026.8kW/5,400hp diesel; 2,043.2kW/2,740hp electric
**Speed:** 20.25 knots (surfaced); 8.75 knots (dived)
**Complement:** 66 men

LEFT: **USS *Pickerel*, one of the Tench class which arrived too late for war service, was among 24 boats selected for the Guppy II conversion and later went into Guppy III mode.** ABOVE: **This is what *Pickerel* could do after that: surfacing at a 48-degree up angle, from a depth of 45.7m/150ft during tests off the coast of Hawaii, March 1, 1952.**

# Tench class

The Tench boats came late to World War II, the first of the class not being laid down until July 1944, but she was speedily built and commissioned on October 6. Even though the build time was very swift indeed, only 31 of a planned 128 boats in this class were commissioned before the end of the war came into view and the remainder were cancelled. Those that did participate came at time when the Japanese were in desperate straits, and had long ago given up any semblance of adherence to the international agreements on the rights and safety of prisoners of war. Consequently, many of the US boats, and indeed surface vessels, were almost daily engaged in the recovery of downed air crews, attempting to snatch them to safety before they could be picked up by enemy ships and submarines.

Allied submarine tactics were by then also well developed, tried and practised, helped by the inclusion of the latest technological aids, particularly in the fields of communication and surveillance. Another development that had successfully been employed by the US submarine command for many months was the utilization of the wolf-pack system of submarine deployment, used to great effect by the Germans. Thus, when *Tench* became the first of her class into battle, she did so in the company of *Balao*, *Sea Devil* and *Grouper* in what were by then described as "co-ordinated attack groups".

The patrol area to which they were assigned encompassed a region of the East China Sea south-west of Kyushu and extended north into the Yellow Sea. Their tasks ranged through rotational patrols, weather reporting, photographic reconnaissance and lifeguard duties. The latter could be extremely hazardous for a submarine, often requiring these massive boats to be manoeuvred close to the shoreline and in the shallows. If trouble came in overhead, there was no place to dive, as *Tench* discovered during her lifeguard stint in support of the 5th Fleet air raids on Nagasaki. Dyes were used to mark the position of downed pilots, and spotter planes kept up a round-the-clock watch. When pilots were spotted, rescue submarines moved in with F6F Hellcat fighters providing cover overhead. On this occasion, the dye was no more than a reflection, and while *Tench* was searching, a flight of bombers appeared out of the sun, but due to the water depth the submarine could not dive. She was a sitting duck. However, the aircraft were friendly and *Tench* escaped unharmed. Further war patrols kept the submarine busy right up to the last, but that was by no means the end of *Tench*'s journey.

In 1950, she and others in her class were converted to Guppy 1A submarines, which involved considerable modifications in almost every department. Thus equipped, she was recommissioned and remained in service until the early 1970s, along with others who had undergone the same modifications.

LEFT: **USS *Argonaut*, also of the wartime Tench class, was reassigned to Submarine Squadron 6 at Norfolk, Virginia. In 1947 she received some of the Guppy modernizations to become one of the first guided missile submarines armed with the new Regulus I missile.**

## Tench class

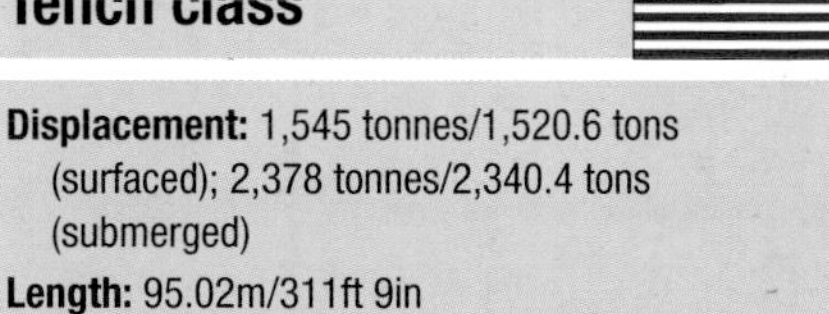

**Displacement:** 1,545 tonnes/1,520.6 tons (surfaced); 2,378 tonnes/2,340.4 tons (submerged)
**Length:** 95.02m/311ft 9in
**Beam:** 8.25m/27ft 1in
**Armament:** 10 x 533mm/21in tubes, 6 x forward, 4 x aft; 24 x torpedoes, 1 x 127mm/5in (25 calibres) deck gun, 1 x 40mm/1.57in gun, 1 x 20mm/0.79in gun, 2 x 12.7mm/0.50 calibre machine-guns
**Propulsion:** 4,026.8kW/5,400hp diesel; 2,043.2kW/2,740hp electric
**Speed:** 20.25 knots (surfaced); 8.75 knots (dived)
**Complement:** 66 men

# Holland

Britain's very first submarine, the *Holland* launched in 1901, can still be seen. She forms part of the collection at the Royal Navy Submarine Museum in Gosport, Hampshire. Bought for £35,000 and constructed under conditions of great secrecy, she was originally known as HM *Submarine Torpedo Boat No. 1*. Initially, Royal Navy engineers and shipbuilders Vickers had great difficult with the drawings, which had many discrepancies, and the British eventually went their own way and severed contact with Holland and the Electric Boat Company of America.

Unlike today's boats with their conning towers and periscopes, *Holland* looked more like a short fat cigar tube with a hatch or scuttle for the captain's head and shoulders, although an experimental periscope was designed and fitted to a later Holland. Built for the Royal Navy under licence, she was a copy of the US Navy's A-class boats.

Steering was by two vertically mounted rudders above and below the single screw and by a horizontal pair of rudders for when the boat was submerged. This design gave anti-submarine sceptics plenty of ammunition, and the boats were certainly no place for anyone with even a hint of claustrophobia. There were no interior bulkheads, very little ventilation and no real accommodation for sitting or sleeping. The fumes were appalling – a combined stench from the petrol engine and the buckets used for ablutions that could only be emptied when they surfaced.

Notoriously unstable, it took almost a year of trials and modifications to settle it down. When surfaced, the boats were difficult to control, and the captain and coxswain had great difficulty hanging on to masts and wheels while on deck, with water washing around their feet. Submerged, the boat seemed to have a constant "nose-down" attitude, which made it appear to be constantly diving. This unnerved the sailors at first, but underwater control and manoeuvrability were described as good.

ABOVE LEFT: **The Royal Navy customized the *Holland* designs they bought from the American Electric Boat Company; consequently, the British and US versions differed somewhat.** ABOVE: **Length and width remained the same, however, and room on top was still scarce.**

Exhaustive trials of *No. 1* lasted three years with the result that *No. 2* became the first submarine to be commissioned into the Royal Navy on August 1, 1903, quickly followed by four more entering service by the end of that year. Despite the experimental environment and their fragility, they saw a decade of service, and many a complacent warship captain was shocked to hear the clang of a practice torpedo hitting his ship's sides. Although only equipped with a single 356mm/14in tube, they were armed with either three long or five short torpedoes. *Holland 1* sank while taking part in manoeuvres off the Nab Lightship in March 1904. All her crew were rescued.

### Holland

**Displacement:** 114.8 tonnes/113 tons (surfaced); 124 tonnes/122 tons (submerged)
**Length:** 19.5m/64ft
**Beam:** 3.58m/11ft 9in
**Armament:** 1 x 533mm/21in tube (bow); 3 x long or 5 x short torpedoes
**Propulsion:** 119.3kW/160hp petrol; 55.9kW/75hp electric
**Speed:** 7.5 knots (surfaced); 5 knots (dived)
**Complement:** 8 men

LEFT: **A pair of *Holland* boats undergoing a clean, probably prior to their display to the public, who were interested in the curious new craft.**

LEFT: ***A4* was one of several boats of this class that came to an unhappy end when the wash from a passing ship sank her, fortunately without loss of life.** BELOW: ***A13* was the survivor of the class, though not with the longevity of HMS *Victory* in the background.**

# A class

Developed from the Holland class, the A-class submarines were 50 per cent longer, but still put to sea with the primitive technology of their predecessors. *A1* was originally intended to be the sixth Holland class. However, with a raft of design improvements on the drawing board it was decided this boat would include the majority of them in an enlarged hull, which was increased by 12.2m/40ft. It was then decided to increase the number of torpedo tubes to two from *A2* onwards and this resulted in a further increase in size by almost 0.6m/2ft and was to become the standard for the rest of the class.

However, the improved design and engine power brought little improvement in terms of performance, with the engine speed changing little from the original Hollands. On the surface, speed was slightly quicker, but submerged things were virtually unchanged. *A5* was the first Group 2 boat and was fitted with a 410kW/550bhp 16-cylinder engine, replacing the 12-cylinder unit, but again this was a disappointment.

Speed increased by only half a knot and the engine was much less reliable. On the surface, this class was still powered by petrol engines, and the dangers soon became apparent. In February 1905, five crewmen on *A5* were killed in an explosion caused by a build-up of petrol fumes after refuelling. *A13* was fitted with an experimental heavy-oil-powered MAN engine built under licence by Vickers – what we call a diesel engine today. Although it used a much safer fuel, it was a much heavier engine, and consequently the boat had to carry a reduced amount of fuel.

It was not until the introduction of the D class that the improved diesel engines replaced the dangerous petrol-powered machinery. However, there were several changes that did work and improve safety – the conning tower being one. It featured two hatches, one leading on to the bridge and a lower one sealing the conning tower from the control room. This prevented water entering the boat from larger waves and enabled it to put to sea in rougher weather. As well as giving better protection to the captain and coxswain when sailing on the surface, it also gave them a better view. *A11* featured an even larger conning tower, and although the lengthened hull gave a better passage than the Hollands while surfaced, overall it was still poor.

As experience was gained it was also found that – using care – the boats could be dived while under way, but with only rudders at the rear, control was difficult. *A7* was therefore fitted with a pair of hydroplanes on her conning tower.

RIGHT: ***A2* was the first boat to be upgraded and slightly elongated to take an additional torpedo tube, and the two tubes fitted side by side.**

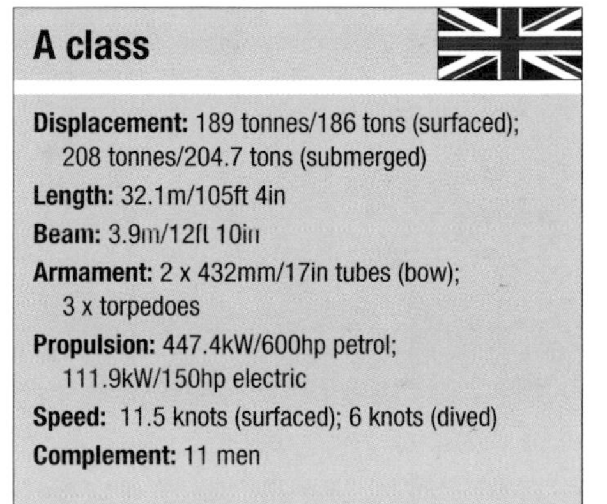

## A class

**Displacement:** 189 tonnes/186 tons (surfaced); 208 tonnes/204.7 tons (submerged)
**Length:** 32.1m/105ft 4in
**Beam:** 3.9m/12ft 10in
**Armament:** 2 x 432mm/17in tubes (bow); 3 x torpedoes
**Propulsion:** 447.4kW/600hp petrol; 111.9kW/150hp electric
**Speed:** 11.5 knots (surfaced); 6 knots (dived)
**Complement:** 11 men

# B class

The B-class boats were the first Royal Navy submarines to have a casing that gave extra protection to the hull and improved buoyancy on the surface. Additionally, they were also the first British submarines to be fitted with a second pair of hydroplanes near the bow in addition to those astern, enabling them to dive more easily when under way. Although surface speed was better than the A class, submerged it remained similar. With the B class came a giant leap forward for the Royal Navy, with the boats ranging further and further from home waters. To patrol in the Mediterranean, they would be towed there by surface ships before deploying. In spite of these long deployments, soon to be in a war environment, there were still no dividing bulkheads, accommodation space or improved ventilation.

Still rather limited in endurance and meant for defensive purposes, the two 432mm/17in torpedo tubes in this class were angled downwards in the belief that torpedoes would be discharged while the submarine rose to the surface. Two reload torpedoes were carried.

ABOVE: ***B4*** **was among the first boats to be taken out of coastal waters and towed to the Mediterranean, opening up vast possibilities for the future.**

Just before Christmas 1914, it was a B-class boat in which Lt Norman Douglas Holbrook became the first submarine commander to be awarded the Victoria Cross, for a daring operation that resulted in the sinking of a Turkish battleship. Holbrook dived *B11* under five rows of mines and attacked the *Messudieh*, returning the same way, and braved attacks by gunfire and torpedo boats. The boat had been submerged for nine hours. As well as Holbrook's VC, all the crew were decorated for gallantry, each man receiving either a DSO or DSM.

One B-class boat was lost with all hands when run down by a surface ship off Dover, and another was sunk by Austrian aircraft. Other boats became inoperable because of a lack of spares and were modified and used as patrol boats instead of submarines, with their conning towers removed and replaced by wheelhouses.

They were also armed with a 12pdr gun. Eleven of these craft were built, and when the war ended only *B3* was operational, largely because she had mainly been used for trials and training.

ABOVE: ***B5*** **demonstrates the relatively unpractised and therefore often hazardous task of taking on torpedoes at sea, which was necessary during operational manoeuvres, especially when far from home. Given that the class carried only two reloads for the two bow tubes, it was a regular occurrence during wartime.**

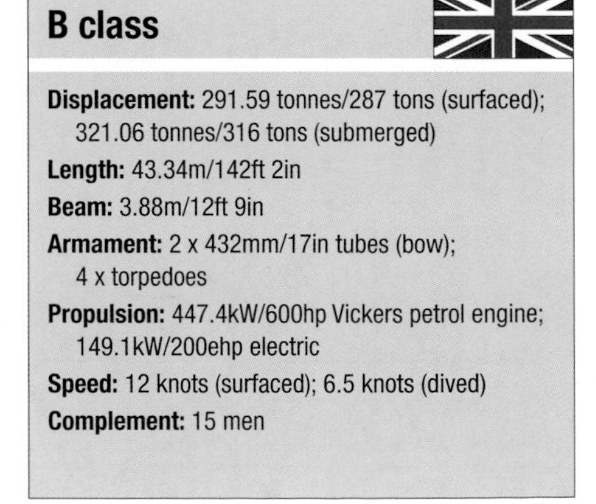

## B class

**Displacement:** 291.59 tonnes/287 tons (surfaced); 321.06 tonnes/316 tons (submerged)
**Length:** 43.34m/142ft 2in
**Beam:** 3.88m/12ft 9in
**Armament:** 2 x 432mm/17in tubes (bow); 4 x torpedoes
**Propulsion:** 447.4kW/600hp Vickers petrol engine; 149.1kW/200ehp electric
**Speed:** 12 knots (surfaced); 6.5 knots (dived)
**Complement:** 15 men

LEFT: **On July 6, 1918, a squadron of German seaplanes returning from a daylight raid on Lowestoft caught Harwich-based *C25* on the surface and used her for target practice. Her captain and six crew were killed outright, but the remaining crew, some badly wounded, managed to close up and dive the boat, and eventually reached the safety of their home base.** BELOW: **An interior view of the C class.**

# C class

Thirty eight C-class boats were built in total, all but six of these by Vickers and for the first time HM Dockyard at Chatham provided submarines for the Royal Navy by building this last half dozen. They also represented the last of the Holland design, and the last to be powered wholly by petrol engines. Given their limited endurance, the C-class boats made a remarkable contribution to the defence of the UK during World War I. All played a gallant and extremely dangerous role, operating mainly in the coastal regions in the North Sea, with many gallantry decorations won by the crews, although losses were also heavy. Safety was also being taken into account, with four of the boats equipped with airlocks and helmets, which were intended to give crew time to breathe extra air before trying to escape through the torpedo hatch.

As the submarine war became more intense, casualties among submariners began to mount and the C boats were to become involved in a daring plan to hit marauding U-boats, which were finding British North Sea trawlers easy pickings. Decoy trawlers would sail into the fishing fleets towing a submerged C boat until a U-boat had been lured into the trap, at which point the submarine was released. *C27* was involved in one such operation, being towed by the decoy trawler *Princess Louise* off Aberdeen, there to attack and sink *U23*. Another successful pairing was that of *C24* and the trawler *Taranaki* to sink *U40*. But then disaster struck. *C33* vanished, presumed victim of a mine while operating with the trawler *Malta*, and *C29* was lost after her towing trawler *Ariadne* strayed into a minefield in the Humber estuary.

Elsewhere, *C7* successfully attacked and sank *U63* and *U68* in 1917, and *C15* disposed of *U65* in what had developed into the dramatic cut and thrust submarine warfare in the North Sea and English Channel. Other British losses included *C31* and *C34*. Four C-class boats were also deployed to the Baltic Flotilla, tasked with attacking German shipping. In 1917, *C26*, *C27*, *C32* and *C35* were loaded on to barges at Archangel and transported to the Baltic using the Siberian Railway. C boats were also included in the taskforce used to block the U-boat base at Bruges, trapping many of the U-boats inside. Two old boats – *C1* and *C3* – were provided for the mission, with the bow of *C3* being filled with explosives and used to block the entrance channel at Zeebrugge.

LEFT: **A typical view of the C-class boats, this one being *C22*, which had an active but uneventful life compared with some of her 37 sister boats. She was among the 24 in the class scrapped at the turn of the 1920s.**

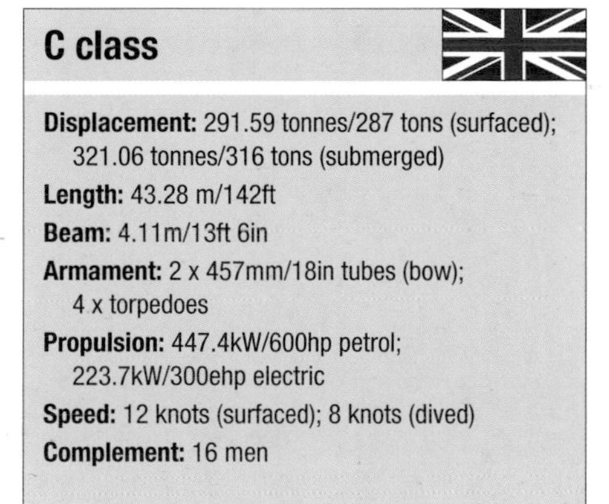

## C class

**Displacement:** 291.59 tonnes/287 tons (surfaced); 321.06 tonnes/316 tons (submerged)
**Length:** 43.28 m/142ft
**Beam:** 4.11m/13ft 6in
**Armament:** 2 x 457mm/18in tubes (bow); 4 x torpedoes
**Propulsion:** 447.4kW/600hp petrol; 223.7kW/300ehp electric
**Speed:** 12 knots (surfaced); 8 knots (dived)
**Complement:** 16 men

# D class

This class saw the arrival of the first diesel-powered submarines of the Royal Navy and the first for which the Admiralty produced the basic design. Twin screws and saddle tanks were introduced along with an enhanced conning tower providing better visibility, and, from *D4* onwards, they were also the first boats equipped with a gun – albeit a paltry 12pdr.

Intended as an effective patrol submarine with overseas potential, the eight D-class boats were meant to undertake patrols in excess of 3,200km/1,988 miles. With better visibility came better manoeuvrability, and the placement of the main water ballast tanks outside the hull in a saddle tank configuration increased space internally. *D4* had its gun in a housing that could be retracted into the conning tower, and it was also fitted with a stern-firing torpedo and a new wireless system that enabled the crew to transmit as well as receive. Mounting the bow torpedo tubes vertically also led to a more streamlined shape. Six of the boats were built by Vickers at Barrow, and the other two at Chatham Dockyard.

At the outbreak of World War I the D boats were stationed along the east coast of Britain, and their role was to support the Grand Fleet and to destroy German warships. *D5* became the first one to fire a torpedo in anger, but the setting used was that for peacetime training and it ran too deep and missed the target. In November 1914 *D3*, *D5* and *E10* were sent to intercept German cruisers bombarding the Norfolk coast, but *D5* was lost after striking a mine.

Shortly afterwards, *D2* was also lost after being hit by a German warship. In 1918, *D3* was sunk accidentally in the English Channel after being bombed by a French airship. *D6* was sunk by a German U-boat off Ireland in 1918, leaving *D7* as the only boat of the class to conduct a successful attack, sinking a U-boat off the north of Ireland in September 1917. Only three boats – *D4*, *D7* and *D8* – survived the war, and they were used for training purposes before being paid off.

ABOVE: **A photograph that puts things into perspective – a D-class boat a few hundred feet from a British battleship. It is easy to see why, in those pioneering days, a submarine could be sunk by the wash of a ship.**

ABOVE: **Launched in May 1910, *D2* became an early victim of World War I when, on November 25, 1914, she was hit and sunk by gunfire from a German patrol vessel in the Ems estuary, The Netherlands.** LEFT: **A profile of the D class.**

**D class** 

**Displacement:** 490.73 tonnes/483 tons (surfaced); 604.52 tonnes/595 tons (submerged)
**Length:** 50.16m/164ft 7in
**Beam:** 6.22m/20ft 5in
**Armament:** 3 x 457mm/18in tubes, (2 x bow, 1 x stern); 6 x torpedoes
**Propulsion:** 2 x 447.4kW/600bhp Vickers diesel engines; 206.6kW/277ehp electric motors
**Speed:** 14 knots (surfaced); 9 knots (dived)
**Complement:** 25 men

LEFT: ***E6*, whose class of 56 submarines had a heroic war, ran out of luck on December 26, 1915, while leaving port for her patrol duty. Nearing a sunken light vessel, she was signalled by a patrolling torpedo boat to keep clear. *E6* continued on her course and within minutes struck a mine and disappeared.**
BELOW: **The torpedo tubes of an E-class boat.**

# E class

The E-class boats were the teeth of the Royal Navy's submarine fleet, carrying out many daring operations throughout World War I and fulfilling both the conventional submarine role and being adapted to work as mine-layers. One was even adapted to carry aircraft in an attempt to counter the threat to London from German airships. Fifty seven of these brilliantly versatile craft were built and 28 were lost. Larger engines were fitted in an attempt to give the boats extra power, and this resulted in an increase in hull size. While a small increase in performance was achieved, the larger hull enabled more fuel and weapons to be carried.

However, with the increase in size came a problem with their manoeuvrability, and it became necessary to position the boat closer to its target in order to fire torpedoes. This gave the enemy an advantage in counter-attack. At one stage, consideration was given to getting rid of the bow torpedo tubes and firing them broadside. Torpedo tubes were introduced in the beam and again the hull had to be increased to accommodate the tubes and loading space.

E class were also the first boats to have watertight bulkheads, which meant compartments could be sealed in emergencies. Previous British submarines had comprised one single compartment. With the bulkheads came other complications in construction, but the boat builders quickly realized the safety benefits and willingly incorporated them in future designs.

Crew comfort was also becoming part of the equation because of the length of time the men were spending aboard. Although it normally took anything from 20 to 30 months to build a submarine, the pressure of war soon began to reduce build-time to around a year, with Vickers completing *E19* in just eight months. This boat went on to become the first Royal Navy submarine to sink a merchant ship, in 1915 in the Baltic, and increased its tally to eight sinkings, including a German light cruiser. *E1* was the first to torpedo a German battleship – attacking the battlecruiser *Moltke* in the Gulf of Riga in August 1915. The boats of the British Baltic submarine flotilla were so successful in their campaign against enemy shipping that the Germans withdrew all major warships from the area.

Six E-class boats were also modified to act as minelayers, with the mines located in chutes in their saddle tanks. An E-class boat – *E3* – also became the first British submarine to be lost in action, when she was attacked by a U-boat in October 1914. Her wreckage was not located until 1997. As well as operating in home waters and the Baltic, the boats carried the war to the enemy in the Dardanelles, braving minefields to carry out their attacks. *E11* sank shipping off Constantinople and sank a Turkish battleship, the *Hayreddin Barbarossa*. Such was the effect of the E boats and the gallantry of their crews that three E boat captains won the Victoria Cross.

*E22* was modified to carry a pair of Sopwith Schneider seaplanes on her deck, and trials began with the aim of launching the aircraft in the North Sea to intercept Zeppelin airships attacking London and other cities. However, the aircraft were too frail to operate in anything other than very calm conditions and the trials were abandoned. Twenty-nine E-class boats survived the war and were finally scrapped between 1921 and 1923. The six minelayers laid almost 2,500 mines.

## E class

**Displacement:** 665.48 tonnes/655 tons (surfaced); 808.73 tonnes/796 tons (submerged)
**Length:** 53.65m/176ft
**Beam:** 6.86m/29ft 1in
**Armament:** 4 x 457mm/18in tubes, (1 x bow, 2 x beam, 1 x stern); 8 x torpedoes; 1 gun of various calibres was carried from 1915
**Propulsion:** 2 x 596.6kW/800bhp Vickers diesel engines; 2 x 313.2kW/420ehp electric motors
**Speed:** 15 knots (surfaced); 9 knots (dived)
**Complement:** 30 men

LEFT: **Alphabetically out of context (using the initial of the manufacturer, Vickers), the early V class was a brave attempt to find the solution to creating the versatile submarine capable of coastal and ocean-going sorties, already being investigated by France, Germany and Japan. In that regard, it was a failure, with a disappointing performance, and the four V boats were relegated to training duties.**

# V class (Early)

Four of the V-class boats were built by Vickers to compete with other experimental boats being constructed at other shipyards. They represented Vickers designers' interpretation of a coastal-type boat designed to meet the requirements of the 1912 Submarine Committee on future submarine development. The innovation was in the double hull section in the middle part of the boat, the result of a number of experimental hulls being tested in a tank.

Vickers built their own diesel engines for the V boats, although the battery was considered small for a boat of its size, and was in two of the class only a third larger than A-class batteries. The estimated cost of the four V-class boats was £75,790 each, and was clearly not the answer to the need for an ocean-going boat that was becoming ever more apparent in December 1913, as Germany pressed ahead with their U-boats. In the meantime, the V boats went into service, but their performance was disappointing, and they lasted no longer than 1919, when they were paid off.

## V class (Early)

**Displacement:** 397.26 tonnes/391 tons (surfaced); 464.31 tonnes/457 tons (submerged)
**Length:** 44.95m/147ft 6in
**Beam:** 4.95m/16ft 3in
**Armament:** 2 x 457mm/18in tubes (bow), 4 x torpedoes; 1 x 12pdr gun
**Propulsion:** 2 x 335.6kW/450bhp Vickers diesel engines; 2 x 141.7kW/190ehp electric motors
**Speed:** 14 knots (surfaced); 9 knots (dived)
**Complement:** 20 men

LEFT: **The pattern of the above V class was utilized for a double-hulled coastal submarine, although with a number of alterations and improvements to the design, including a stern torpedo tube, with no great success, either.**

# F class (Early)

The F class were to be compared with other experimental boats, based on the V class and had similar performance. Although said to be less stable, *F1* and *F3* were equipped with Vickers diesel engines, with a MAN diesel engine installed in the third boat for comparison.

They were also fitted with an additional stern-mounted torpedo tube. The three were built between 1913 and 1917 and, at one time, there were plans for five of them to be built. This was scrapped once war broke out. They were used for local defence patrols and later sent to Portsmouth and Campbeltown to be used for training before being paid off in 1919.

## F class (Early)

**Displacement:** 368.8 tonnes/363 tons (surfaced); 448.06 tonnes/441 tons (submerged)
**Length:** 46.02 m/151ft
**Beam:** 4.88m/16ft
**Armament:** 3 x 457mm/18in tubes (2 x bow, 1 x stern), 6 x torpedoes
**Propulsion:** 2 x 335.6kW/450bhp Vickers diesel engines; 2 x 149.1kW/200ehp electric motors
**Speed:** 14.5knots (surfaced); 8.75 knots (dived)
**Complement:** 19 men

LEFT: **The *G13* immediately became part of the Royal Navy's 10th Flotilla, based in the Tees, in a class which was the first to be armed with 533mm/21in torpedoes, with which *G13* sank the German *UC43* off Muggle Flugga light house on March 10, 1918.** BELOW: **The G-class profile, based on an Italian double-hull design.**

# G class

The G-class boats, of which 13 were built, were supposed to become the British answer to ocean-going submarines to match the U-boats, but didn't quite make it. The design features added to the continuing improvements in overall accommodation, allowing more and better equipment to be installed; however, they cost £125,000 apiece. The crew were better off, with improved living conditions that even included an electric oven. The two eight-cylinder diesels generated a worthwhile 1,193kW/1,600bhp, providing a surface speed of 14 knots, but undoubtedly the biggest improvement brought about by early war experience was the upgrade in armaments, which saw the beginning of larger torpedoes as standard. The specification was changed to provide two 457mm/18in bow tubes, two 457mm/18in beam tubes and one 533mm/21in stern tube.

The class also carried one 76mm/3in Quick Firing High Angle (QF HA) gun, which was fitted just forward of the bridge, and a portable 2pdr fixed to a pedestal at the after end of the bridge. In operations the boat is also known for diving to what was described as an exceptional depth of 51.8m/170ft, when being chased in error by three British destroyers.

During World War I, *G7*, *G8* and *G11* were lost on active service through unknown causes, and in September 1917, HMS *Petard* sank *G9* in error off the Norwegian coast after the submarine had attacked the ship – also in error. Of the boats that survived, four were taken out of service at the end of the war, and the remaining six were withdrawn from service in January 1921. They had proved something of a disappointment to many, given that their performance was little better than that of the E class, even with the incorporation of the double-hull concept.

LEFT: **The control compartment of the G-class boats demonstrated the rapid pace of improvements aboard the new submarines coming on stream, as beam and length were increasing with each new design.**

## G class

**Displacement:** 710.18 tonnes/699 tons (surfaced); 982.47 tonnes/967.9 tons (submerged)
**Length:** 57.02m/187ft
**Beam:** 6.86m/22ft 6in
**Armament:** 4 x 457mm/18in tubes (2 x bow, 2 x beam); 8 x torpedoes; 1 x 533mm/21in tube (stern); 3 x torpedoes; 1 x 76mm/3in gun
**Propulsion:** 2 x 596.6kW/800bhp Vickers diesel engines; 2 x 313.2kW/420ehp electric motors
**Speed:** 15 knots (surfaced); 9 knots (dived)
**Complement:** 30 men

LEFT: **The stylish-looking H class, rushed into service for World War I, was of an American design, with some being built in Britain and Canada. They proved a handy boat with the fastest dive, down in thirty seconds.** ABOVE: ***H23*** **photographed in the Kiel Canal.**

# H class

As World War I continued to put pressure on British industry to feed the war machine, the country's shipyards were unable to satisfy the Admiralty's demands for more submarines. The Admiralty despatched a delegation to the United States, and 20 single-hulled craft were ordered from the Electric Boat Company.

The boats were to be shipped to the UK for assembly as America had not yet entered the war, but this idea was to fall foul of the US government's neutrality laws. It was then decided to build the first ten at Vickers Canada, with another ten being built by the Electric Boat Company. These, however, were interned at Boston by officials, still fiercely guarding America's neutrality while at the same time selling boatloads of vital supplies to Germany.

Eventually they were released to the Royal Navy, but by then the Canadians had shown the way by sending their boats across the Atlantic – the first such voyage by a submarine – and were in service in 1915. The first four, escorted by a warship, went straight on deployment in the Mediterranean, and the other six sailed to Britain escorted by a cruiser and support freighters.

With their up-to-date accommodation and facilities, these boats were very popular with their crews, and several of them were still serving in various navies in World War II. So successful was the H class that a further batch was ordered. They were to be built by Vickers at Barrow and were to be armed with the bigger 533mm/21in torpedo and fitted with a more powerful radio. Most of the components were shipped from America. A further 22 boats were ordered from various British shipbuilders in 1917, and most were delivered in 1918–19.

This completed a total fleet of 54 H boats in three groups, which until that point in time was only exceeded by the E class, although a number were not completed in time for action, and of those, eight American-built boats were ceded to the Canadian and Chilean Navy, the latter still running them in 1953 before they were scrapped.

LEFT: ***H31*** **was not completed in time for World War I, but was immediately in action in World War II and sank a U-boat on July 18, 1940. In December, she was involved in operations against the German battleship *Scharnhorst* near Brest, from which she failed to return, assumed mined.**

**H class**

**Displacement:** 369.82 tonnes/364 tons (surfaced); 440.94 tonnes/434 tons (submerged)
**Length:** 45.79m/150ft 3in
**Beam:** 4.80m/15ft 9in
**Armament:** 4 x 457mm/18in torpedo tubes (bow); 8 x torpedoes
**Propulsion:** 313.2kW/420hp diesel; 231.2kW/310ehp electric
**Speed:** 13 knots (surfaced); 10 knots (dived)
**Complement:** 22 men

# J class

The J-class submarine came into being because the Admiralty wanted a craft capable of sailing with the surface ships of the Grand Fleet at a speed of 21 knots on the surface. It was not to be achieved. Because of the extra power needed, the boat's length was increased by more than 50 per cent over the E class that it was intended to replace. An extra shaft and third diesel engine would be needed to try to increase the power. However, all this achieved was a bigger and heavier hull with an increase in speed of just four knots, bringing it up to 19 knots on the surface.

Once this became apparent the Admiralty turned their attention to the even larger K class, which was also under construction, and the J-class boats were reclassified as overseas boats. Eight J-class boats had been ordered in January 1915, but by April two had been cancelled. Another boat had been ordered with a modified design but, again, there was no great improvement in performance.

*J7* had a different layout to others of the class, with the control room positioned between the two engine rooms and the conning tower built slightly further back, giving better access space for the bow torpedo tubes.

They operated from their base in Blyth with the Grand Fleet, but because of their insufficient performance saw little action. However, *J1* attacked four German warships with a salvo of torpedoes, seriously damaging the battleships *Kronprinz* and *Grosser Kurfurst*. At the end of World War I the six remaining J-class boats were transferred to the Royal Australian Navy.

TOP: **Most of the J-class boats were in action by mid-1916. *J1* had a spectacular initiation under Lt Commander Laurence when, sighting four battleships off Horns Reef, he fired four torpedoes and hit two enemy dreadnoughts.** ABOVE: **A profile of a J boat.** LEFT: ***J6* came to an unhappy end due to friendly fire, having been mistaken for *U6* and sunk by gunfire from the Q ship *Cymric* off Blyth on October 15, 1918. Fifteen survivors were picked up, but 19 other crewmen were killed.**

### J class

**Displacement:** 1,223.26 tonnes/1,204 tons (surfaced); 1,849.12 tonnes/1,820 tons (submerged)
**Length:** 83.74m/274ft 9in
**Beam:** 7.16m/23ft 6in
**Armament:** 6 x 457mm/18in torpedo tubes (4 x bow, 2 x beam); 12 x torpedoes; 1 x 76mm/3in gun,1 x 12pdr gun, and depth charges
**Propulsion:** 3 x 894.8 kW/1,200bhp Vickers diesel engines; 3 x 335.6kW/450ehp electric motors
**Speed:** 19 knots (surfaced); 9.6 knots (dived)
**Complement:** 44 men

# Nautilus

This was the first Royal Navy submarine to be given a name, and in retrospect can be seen as a stepping stone to larger submarines. Designed with a double hull and increase in length, *Nautilus*' sea-keeping qualities, particularly in bad weather, were a great improvement on other submarines. Laid down in March 1913, she had an overall length of almost 79.2m/260ft, and was twice the size of any existing submarine. She was also the most expensive boat hitherto built by the Royal Navy, with costs escalating to almost £250,000.

Even so, there were a number of disappointing factors, especially in her speed, which did not meet the 17 knots on the surface called for in the design specifications. Her weaponry comprised two 457mm/18in bow torpedo tubes, four 457mm/18in beam tubes and two 457mm/18in stern tubes, with a combined load of 16 torpedoes carried. A 76mm/3in High-Angle gun was fitted on the superstructure just forward of the bridge, and this was raised and lowered on a vertical ram.

In the event, *Nautilus* never saw any active service and was used largely as a depot ship and for training purposes. However, there is no doubt that she provided a link in the British shipbuilding industry between the smaller boats and the great developments that followed in the quest for larger submarines.

The Admiralty was spurred on to seek better performances than 17 knots, and although engine performance was generally improving, as engines almost doubled in power during the building of these experimental boats, many difficulties and teething problems lay ahead. As for *Nautilus*, she was paid off in 1919 and sent to the scrap yard three years later.

ABOVE: ***Nautilus*, a large experimental boat built during World War I and described by the Royal Navy as an "exceedingly interesting" project, was ultimately a failure. She ended her days charging batteries for other submarines.**

### Nautilus

**Displacement:** 1,464.06 tonnes/1,441 tons (surfaced); 2,058.42 tonnes/2,026 tons (submerged)
**Length:** 78.73m/258ft 4in
**Beam:** 7.92m/26ft
**Armament:** 8 x 457mm/18in tubes (2x bow, 4 x beam, 2 x stern), 16 x torpedoes; 1 x 76mm/3in deck gun
**Propulsion:** 2 x 1379.5kW/1,850bhp Vickers diesel engines; 2 x 372.8kW/500ehp electric motors
**Speed:** 17 knots (surfaced); 9 knots (dived)
**Complement:** 42 men

LEFT: **The first of two boats named *Swordfish*. This one, built in 1916 (the other came with the 1930s S class), was an experimental steam turbine-powered submarine designed for long-range work, a development from the J class and the predecessor of the steam-powered K class.**

# Swordfish

*Swordfish* was built as part of the Admiralty's ongoing attempt to have submarines which could keep up with the Grand Fleet on the surface at a speed of 21 knots, which was to be achieved by installing a remarkable steam-powered propulsion system, supposedly making her much safer. However, it became very hot inside.

During her trials it was discovered that despite having a mount which meant the gun could be stowed, an exposed gun created less water resistance than had been estimated in theory. The biggest disappointment was that the speed was hardly improved upon, and she had poor manoeuvrability, a dangerous failing when operating in company with the Grand Fleet. Eventually *Swordfish* was converted into a surface patrol boat and fitted with two 12pdr guns.

### Swordfish

**Displacement:** 946.91 tonnes/932 tons (surfaced); 1,493.52 tonnes/1,470 tons (submerged)
**Length:** 70.48m/231ft 3in
**Beam:** 6.93m/22ft 9in
**Armament:** 2 x 533mm/21in tubes (bow), 4 x torpedoes; 4 x 457mm/18in tubes (beam), 8 x torpedoes; 2 x 76mm/3in guns
**Propulsion:** 2 x 1,491.4kW/2,000shp steam turbines; 2 x 1,044kW/1,400ehp electric motors
**Speed:** 18 knots (surfaced); 10 knots (dived)
**Complement:** 42 men

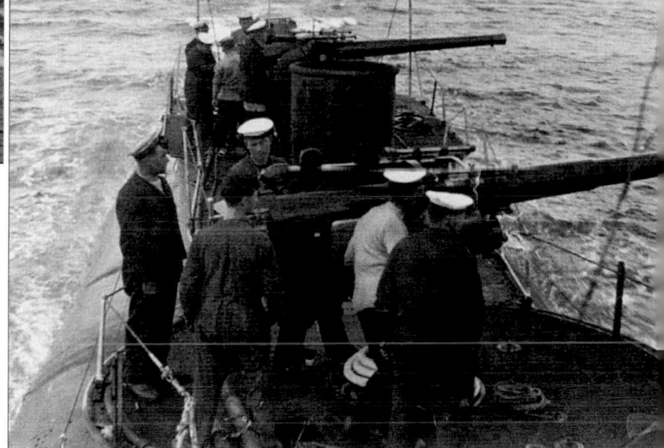

LEFT: *K26* was a new and improved version of the K class completed in 1923, but no others were built. She was noted for completing a successful journey to the Far East via Malta, through the Red Sea to Colombo, Singapore and back.
BELOW: The gunnery tasks on the K class become something of a nightmare, given the length of the boat, with ammunition stored in the bow and stern, thus requiring a long-distance haul by manpower.

# K class

The obsession in certain quarters of the Admiralty for a fleet submarine that could operate with the Grand Fleet led to the K class. Designed to travel at 24 knots, it was also a reaction to a rumour around 1914 that the Germans had secretly developed a boat capable of 22 knots. With the current crop of petrol and diesel engines incapable of delivering the power required to drive a boat at that speed, steam power was chosen. As a production class, these were the largest and fastest submarines in the world with firepower to match, and such was the prestige surrounding them that ambitious officers cherished command of one of these boats – until the reality of sailing them hit home.

Initially they were armed with six torpedo tubes in the bow, four in the beam and three guns – but two of the bow tubes soon had to be removed to improve stability, followed by one of the guns. Even with a speed comparable to that of a destroyer, their manoeuvrability was a problem. Their turning circle was more like that of a dreadnought. They were also slow to submerge and once under water difficult to control. Gunnery, too, was a problem as the guns were slow to bring to bear because of poor control and the ammunition being stored in the bow or the stern requiring a lot of men to get the shells to the guns.

Diving could take anything from three to five minutes because of the number of hatches, vents and other equipment which needed to be closed before the 28 ballast tanks could be blown – and this was another operation fraught with danger, to ensure the boat did not end up in a nosedive or submerge with a list. Forty valves had to be operated in a carefully-controlled manner.

It was also discovered that because of the speed at which the boats travelled on the surface with the Grand Fleet, the door shutters of the torpedo tubes were often unable to survive the high water pressure, which left the tubes fouled and inoperable. The crews who operated these boats became increasingly alarmed about their safety because of what they regarded as the vessels' poor seaworthiness.

Their forebodings were to prove correct as 16 major accidents and eight disasters hit the K-class fleet, costing lives and boats. On only one occasion did a K-class boat fire its torpedoes at the enemy when engaging a U-boat, and even then it missed.

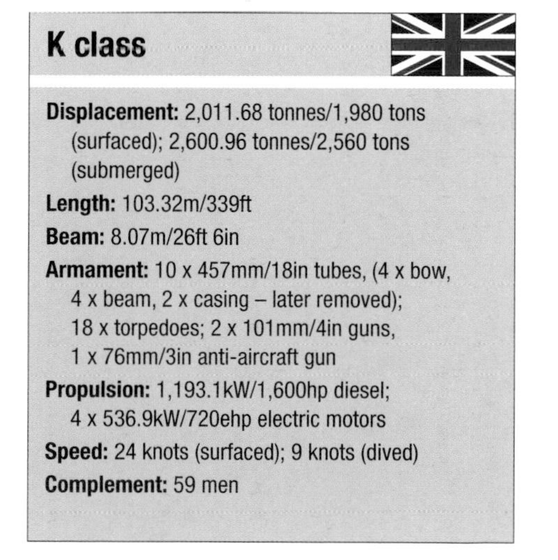

**K class**

**Displacement:** 2,011.68 tonnes/1,980 tons (surfaced); 2,600.96 tonnes/2,560 tons (submerged)
**Length:** 103.32m/339ft
**Beam:** 8.07m/26ft 6in
**Armament:** 10 x 457mm/18in tubes, (4 x bow, 4 x beam, 2 x casing – later removed); 18 x torpedoes; 2 x 101mm/4in guns, 1 x 76mm/3in anti-aircraft gun
**Propulsion:** 1,193.1kW/1,600hp diesel; 4 x 536.9kW/720ehp electric motors
**Speed:** 24 knots (surfaced); 9 knots (dived)
**Complement:** 59 men

# L class

Unlike the K-class boats, the first batch of eight L class was developed from the successful E class, had increased speed, better range, and improved armament. They were equipped with six 457mm/18in torpedo tubes, four forward and two beam firing. A second batch of 26 L-class boats was then ordered, but only 18 of them were actually built. These were armed with four 533mm/21in torpedo tubes in the bow, which also necessitated lengthening of the boat. Six were then converted to minelayers, losing their beam tubes and replacing them with vertical chutes capable of launching 16 mines that were located in the saddle tanks.

A further 25 were then ordered, although only seven were delivered. On these craft, the number of bow tubes was increased to six and the beam tubes were done away with altogether. Nearly all were fitted with a 101mm/4in gun, and the last batch had a rear-firing gun located in an enlarged conning tower. In May 1926, during the General Strike, *L12* was one of several submarines used as an emergency electricity generator moored at the Royal Victoria and Albert Docks.

Popular with their crews, one of them – *L8* – became known as "The Boat That Would Not Sink". Having survived two collisions in 1918, one of them with another L craft, she was towed into the Channel as a target in 1929 after being paid off when trials of new 119mm/4.7in ammunition were being carried out. After being hit several times, she remained afloat and was towed back to Portsmouth for further inspection before being finally scrapped. Although obsolete by the outbreak of World War II, L boats initially served operationally before being transferred to the training role and then taken to Canada to give anti-submarine training to convoy escorts. The last operational boat of this class, *L23*, was finally paid off in October 1945, and foundered off Nova Scotia while being towed to the breakers yard.

TOP: **The first of the L class were among the last to be commissioned in World War I, but they were a popular utility boat that continued on into the 1930s, although most were disposed of by the outbreak of war.** ABOVE: **The profile of *L23*, launched in 1919.**

LEFT: **The firepower varied on the L class, and although they were designed to carry two 76.2mm/3in D/HA guns, most were only fitted with one.**

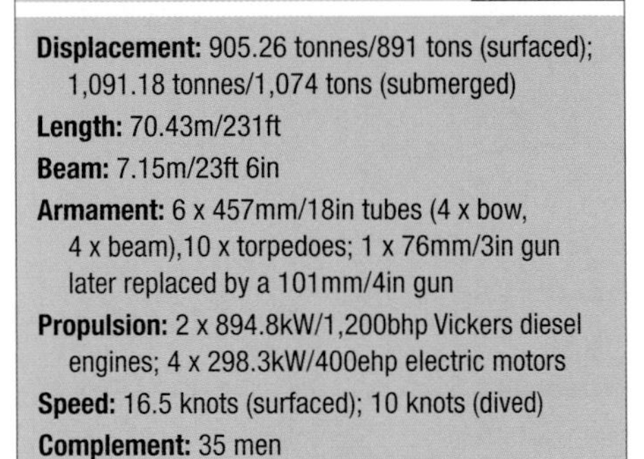

**L class**

**Displacement:** 905.26 tonnes/891 tons (surfaced); 1,091.18 tonnes/1,074 tons (submerged)
**Length:** 70.43m/231ft
**Beam:** 7.15m/23ft 6in
**Armament:** 6 x 457mm/18in tubes (4 x bow, 4 x beam),10 x torpedoes; 1 x 76mm/3in gun later replaced by a 101mm/4in gun
**Propulsion:** 2 x 894.8kW/1,200bhp Vickers diesel engines; 4 x 298.3kW/400ehp electric motors
**Speed:** 16.5 knots (surfaced); 10 knots (dived)
**Complement:** 35 men

LEFT: ***M2*, originally built with a 305mm/12in gun, was converted to carry aircraft by having a hangar installed in the empty space. A gantry was fitted to lift a Parnell Peto spotter plane. But disaster struck when the hangar doors were left open, and the submarine sank, with the loss of 60 men.** ABOVE: **The profile of the M class with the gun fitted.**

# M class

Admiral Sir John "Jackie" Fisher, the man who conceived the Dreadnought class of battleships, came up with the idea of a dreadnought submarine. He had little faith in the torpedoes of the day, believing they were incapable of stopping, let alone sinking, a warship.

These submarines would have a 305mm/12in gun forward of the conning tower similar to those fitted to the Majestic battleships. A smaller 76mm/3in gun would also be fitted as an anti-aircraft weapon. The main gun weighed 123.95 tonnes/122 tons, and 40 rounds – weighing a further 29.5 tonnes/29 tons – were stowed. Then there was the gun crew – 11 specialist gunnery ratings plus 16 in the ammunition party for the main gun, plus another six for the anti-aircraft gun. One unexpected bonus was that all the additional weight made the boat very stable.

The gun mount had a non-watertight housing, and the mechanism for loading the shell and charge were inside a watertight tower. The gun had a range of 29,900m/32,699yds, and the boat had to surface to load and fire. Accuracy would be approximate because of the movement of the boat on the sea and the lack of an artillery spotter.

Approaching a target – whether afloat or on the enemy coast – the boat would pop up and down from periscope depth until it got into position. Then it could surface and fire between 30 and 40 seconds before disappearing back to periscope depth in a further 50 seconds. Although Admiral Fisher came up with the proposal in 1915, none of the three M-class boats was to fire its armament in anger. *M1* did not go to sea until 1919 and was joined by *M2* the following year. Although the gun was waterproofed at the muzzle end, it was not made watertight at the breech. Water in the barrel when firing was to prove disastrous, and at least four barrels were burst because of it.

The Washington Disarmament Treaty of 1920 meant that no submarine could have a gun larger than 203mm/8in, so *M2* and *M3* had their guns removed, with *M2* being converted to an aircraft carrier with a hangar constructed where the gun mount had been. She carried a single Parnell Peto aircraft specially designed for the project. She became the first submarine to have a watertight hangar which could be sealed when she was submerged. The boat was able to surface and launch its aircraft in around 12 minutes. However, in January 1932, one of the hatch doors was left open and she sank off Portland Bill, with the loss of her entire crew.

RIGHT: **The attractive looking *M1*, leader of the class of three converted K boats was rammed and sunk by a Swedish collier in the English Channel in November 1925, with the loss of 69 crew.**

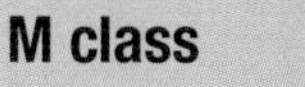

## M class

**Displacement:** 1,635.76 tonnes/1,610 tons (surfaced); 1,977.13 tonnes/1,946 tons (submerged)
**Length:** 90.14m/295ft 9in
**Beam:** 7.49m/25ft 7in
**Armament:** 4 x 483mm/19in tubes (bow), 10 x torpedoes; 1 x 305mm/12in gun, 1 x 76mm/3in anti-aircraft gun
**Propulsion:** 2 x 894.8kW/1,200bhp Vickers diesel engines; 4 x 596.6kW/800ehp electric motors
**Speed:** 15 knots (surfaced); 9 knots (dived)
**Complement:** 64 men

LEFT: **With a streamlined shape, the first R-class boats were built for speed underwater, and this was achieved. However, there were setbacks in other areas of production, and although ten were built at considerable expense, all were scrapped within eight years.**

### R class (Early)

**Displacement:** 421.64 tonnes/415 tons (surfaced); 513.08 tonnes/505 tons (submerged)
**Length:** 49.91m/163ft 9in
**Beam:** 4.61m/15ft 1in
**Armament:** 6 x 457mm/18in tubes (bow), 12 x torpedoes
**Propulsion:** 1 x 179kW/240bhp American diesel engine; 2 x 894.8kW/1,200bhp electric motors
**Speed:** 10 knots (surfaced); 15 knots (dived)
**Complement:** 22 men

# R class (Early)

These boats were the first anti-submarine submarines, or what today are called hunter-killers. Entering service with the Royal Navy in 1917 they were designed to be quicker submerged than when surfaced. This they achieved with a speed of 15 knots that was not to be bettered until the latter stages of World War II, more than a quarter of a century later. The R class had six 457mm/18in torpedo tubes in the bow, and although it had originally been intended to mount a 101mm/4in gun, this was omitted in order to keep the weight down. The superstructure was as light as possible and ballast tanks were internal to keep the shape streamlined.

This boat also had problems. Although fitted with two powerful 895kW/1,200bhp electric motors, which enabled it to reach 15 knots when submerged, the 179kW/240bhp diesel engine was not strong enough to charge the 220 batteries, and because of this lack of power, it was a difficult craft to steer on the surface. In addition to the periscope, the boats had five primitive hydrophones to assist with detecting enemy submarines.

The boats only ever mounted one attack on a German U-boat, and the torpedo failed to explode. Of the 12 ordered, 10 were built and all were paid off by 1925.

LEFT: **A profile of the twin-turreted *X1* cruiser submarine with six torpedo tubes – a brilliant-looking boat that promised great potential, but was a disappointment in the end.**

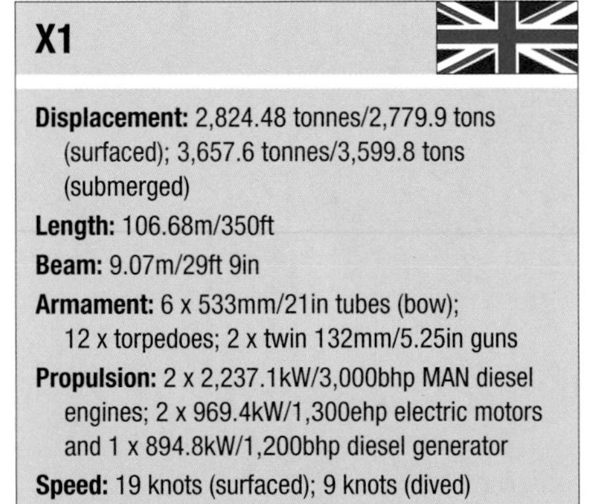

### X1

**Displacement:** 2,824.48 tonnes/2,779.9 tons (surfaced); 3,657.6 tonnes/3,599.8 tons (submerged)
**Length:** 106.68m/350ft
**Beam:** 9.07m/29ft 9in
**Armament:** 6 x 533mm/21in tubes (bow); 12 x torpedoes; 2 x twin 132mm/5.25in guns
**Propulsion:** 2 x 2,237.1kW/3,000bhp MAN diesel engines; 2 x 969.4kW/1,300ehp electric motors and 1 x 894.8kW/1,200bhp diesel generator
**Speed:** 19 knots (surfaced); 9 knots (dived)
**Complement:** 109 men

# X1 class

Only one X1-class boat was ever built – the biggest submarine in the Royal Navy until the nuclear boats came into service decades later. Although conceived by the Royal Navy Submarine Committee in 1915 as a cruiser submarine armed with big guns as well as torpedoes, she was not laid down until November 1921. Dyed-in-the-wool naval planners still wanted submarines operating as part of the Grand Fleet, but submarine warfare experts wanted her to be able to operate in an independent role – a large, fast boat operating at extended ranges. At 2,824 tonnes/2,779 tons on the surface, she was certainly that. After World War I the navy had time and opportunity to carefully examine captured German submarines, including their cruiser submarines. She would also be the first new boat since the war.

*X1* was completed in September 1925, joined the Navy the following spring and was the largest submarine in the world at that time. She was sent on a return trip to Gibraltar but problems arose with her diesel engines, and she was sent back to the RN Dockyard at Chatham for repairs.

The following January she passed all her tests and was assigned to the Mediterranean Fleet, but within a year she again had engine trouble, this time in Malta. There were other niggles, too. Officers claimed that she handled well and that most of her problems were because of crew errors. *X1* was laid up in reserve in 1933 and scrapped three years later – the only Royal Navy submarine built after World War I that did not survive until World War II.

# O class

These were long-range submarines based on the L class and designed to operate in Far Eastern waters as concerns grew about Japanese intentions after failure to renew the Anglo–Japanese Alliance in 1922. The first one became *Oberon*, and although 2 knots slower than the L class, the hull was 12m/39ft 4in longer and they could carry double the number of torpedoes. Their range was also nearly double that of their predecessor. In addition, their wireless range was better and diving depth was deeper. The 101mm/4in gun mounted on the front of the conning tower was later modified to be fitted on a revolving mount. *Oberon* was also the first boat to be fitted with the ASDIC submarine detection system, which evolved into today's sonar.

An O-class boat, *Oxley*, became the first Royal Navy submarine to be lost in World War II when another submarine, *Triton*, mistakenly torpedoed her off the Norwegian coast in 1939. It was only when two survivors were plucked from the water that the true horror unfolded. *Triton* had been signalling *Oxley*, but when no recognition signal was received, assumed her to be an enemy U-boat and torpedoed her. With the recovery of the crewmen, they discovered *Oxley* had been returning the signals but her signal lamp was faulty. A full-scale cover-up was ordered, and it was not until the 1950s that the true account of what happened was released. At the time, a cover story was put out that she had suffered an accidental explosion.

*Oberon* and *Otway* both went on operations in World War II, but were later transferred to a training role. A second batch of six O-class boats were fitted with a more powerful diesel engine that gave them a higher surface speed, and their hydroplanes were moved from the bottom of the pressure hull to the top and hinged so they could be turned in when not in use. Although this increased dive time, it eradicated a lot of the damage they suffered when the boat was submerged for long periods of time.

As the Germans tightened their stranglehold on the strategic Mediterranean island of Malta by sea and air, causing havoc among the convoys trying to supply the beleaguered garrison, the O-class boats – the first to be given names instead of numbers – joined other large submarines to run the gauntlet underwater by ferrying in vital supplies of ammunition, aviation fuel, torpedoes, food and fresh water.

ABOVE: ***Otway*, *Oberon* and *Oxley*, a trio of O-class boats built in two groups from the late 1920s, including two that joined the Australian Royal Navy but which returned for wartime duties.** LEFT: **Chile later bought and modified O-class boats, and renamed them as O'Brien class.**

LEFT: ***Oberon*, the class leader commissioned in 1927, was rather troublesome, suffering breakdowns that forced her into reserve status until she was brought out for war duty and then scrapped. She gave her name to the Oberon class of the 1960s.**

## O class

**Displacement:** 1,513.84 tonnes/1,489.9 tons (surfaced); 1,922.27 tonnes/1,891.9 tons (submerged)
**Length:** 82.19m/269ft 8in
**Beam:** 8.53m/28ft
**Armament:** 8 x 533mm/21in tubes (6 x bow, 2 x stern), 16 x torpedoes;1 x 101mm/4in gun
**Propulsion:** 2 x 1,006.7kW/1,350bhp Admiralty diesel engines; 2 x 484.7kW/650ehp electric motors
**Speed:** 13.5 knots (surfaced); 7.5 knots (dived)
**Complement:** 53 men

LEFT: ***Pandora*** **was engaged in running supplies to Malta and was unloading when an enemy bombing raid began. Her crew courageously carried on only to receive two direct hits from bombs, and she sank on April 1, 1942.** BELOW: **HMS *Perseus* was mined off Cephallonia, leaving a sole survivor from the crew of 51 to tell the tale.**

# Parthian class

Like the O class before them, boats of the Parthian class were prone to leaking fuel from their saddle tanks when submerged, which then left a giveaway smear across the ocean's surface. The leaks were caused by the pressure on the riveted tanks. One of the boats, *Poseidon*, was lost after colliding with a Chinese coaster in the Yellow Sea in June 1931, but 35 of the 57-man crew managed to survive by taking part in the first big escape from a submarine using the Davis Submarine Escape Apparatus.

During the siege of Malta in 1941–42, *Parthian*-class boats *Pandora* and *Parthian* along with other submarines were used as supply craft. A section of their batteries was removed and their spare torpedoes left behind to create additional space for stores for the garrison to be carried. Petrol and other fuels were also ferried in, and fresh water was stored in the ballast tanks.

One of the most remarkable escape and survival stories of World War II involved *Perseus*, which struck an Italian mine in the Ionian Sea. Leading Stoker John Capes and four other survivors were trapped in the submarine at a depth of 52m/171ft. Their only means of escape was the Davis Apparatus. After downing a bottle of rum for courage, they began their escape attempt, but only Capes made it to the surface alive. Exhausted, he managed to struggle a further 8–10km/5–6 miles to the island of Cephalonia, where he was found by locals who looked after him before handing him over to the Resistance. It took 18 months for him to reach home, and he was awarded the British Empire Medal. There were sceptics who did not accept his account of his escape. It was not until 1998 that the wreck of the *Perseus* was located with all the evidence, including the empty rum bottle. Capes went on to become a Chief Petty Officer, but, sadly, he died ten years before the wreck was found and his story verified.

One of the most deadly of the *Parthian*-class boats was *Proteus*, which had spent the pre-war years in the Far East. Deployed to the Mediterranean with other large submarines to help Malta, she was eventually sent back to Britain in 1943, but not before she had damaged or sunk 11 Italian transports.

LEFT: ***Parthian*** **led the P class to China and then to the Mediterranean when war broke out. She is known to have sunk two Italian submarines, the Vichy French submarine *Souffleur* and other Italian vessels before she was herself posted lost, presumed mined in 1943.**

## Parthian class

**Displacement:** 1,796.29 tonnes/1,767.9 tons (surfaced); 2,067.56 tonnes/2,034.9 tons (submerged)
**Length:** 88.13m/289ft 2in
**Beam:** 9.12m/29ft 11in
**Armament:** 8 x 533mm/21in tubes (6 x bow, 2 x stern), 14 x torpedoes;1 x 101mm/4in gun
**Propulsion:** 2 x 1,730kW/2,320bhp Admiralty diesel engines; 2 x 492.1kW/660ehp electric motors
**Speed:** 18 knots (surfaced); 8.5 knots (dived)
**Complement:** 53 men

LEFT: **HMS *Scorcher* was a late arrival in the large class of more than 60 S-class boats. Classic workhorses of World War II, although many were not commissioned until the latter stages, some, like *Scorcher*, remained in service for 20 years or more.** BELOW: **A profile of Group III S-class boat, *Storm*.**

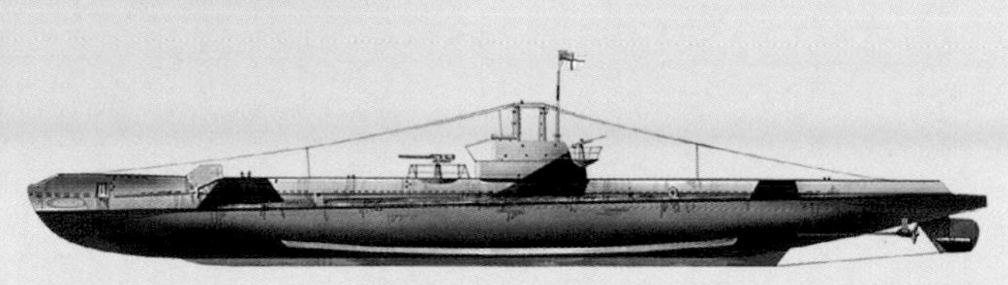

# S class

The S boats carried out some of the most daring and dangerous and, at times, most celebrated operations of World War II. With an increased capability in speed, range and depth, they were quicker to dive and were also the first boats to have escape hatches located both fore and aft, something which was to save the lives of many submariners as the war grew in intensity and they operated in some very hostile environments.

Slightly smaller than some other classes, they were able to operate in shallower and more confined waters. The first four boats joined the fleet in 1932–33 and were equipped with a 76mm/3in gun positioned in a forward extension of the conning tower that was stowed before diving. A second batch of eight boats was ordered with the gun mounted on the casing, which meant the conning tower could be reduced in height, although a small extension to the front served as an ammunition bunker. The next tranche of these craft began production at the outbreak of the war, and production continued throughout, with variations to lines and equipment as innovations were discovered. Eventually, instead of being riveted or part-riveted, the hulls were all welded. Air-warning radio direction finders, ASDIC, external stern tubes and more powerful anti-aircraft guns were introduced.

In addition to the German threat, these submarines suffered so-called "friendly fire" attacks from RAF aircraft, other submarines and, occasionally, Royal Navy ships. A lot of the British torpedoes were also found to be faulty, a problem shared by the German U-boat packs. An S-class submarine, *Sturgeon*, carried out the first successful Royal Navy submarine attack of the war when she sank the anti-submarine trawler *V209* off Heligoland in November 1939. In addition, *Salmon* carried out the first successful attack on a U-boat, sinking *U36* three weeks later. Just over a week later, she attacked a formation of German warships, hitting the light cruisers *Leipzig* and *Nurnberg*. *Spearfish* attacked and damaged the German pocket battleship *Lützow*, leaving her helpless with her propellers and steering gear out of action. *Lützow* had to be towed to port and was out of action for a year. However, the success was not one-sided and many brave men were lost with their boats after clashes with U-boats, aircraft and surface craft. Others perished when their boats struck mines.

*Seraph* was certainly the most famous of these boats, particularly after the war, with two films made about her exploits: the first, *The Man Who Never Was*, and the second, *The Ship With Two Captains*. The boats were also used to gather intelligence about possible landing sites for various operations to liberate Europe from the Germans.

RIGHT: **A quartet of S-class boats, *Shark*, *Sunfish*, *Snapper* and *Sealion*, of whom sadly only the latter survived the war.**

## S class Group 1

**Displacement:** 650.24 tonnes/640 tons (surfaced); 941.83 tonnes/927 tons (submerged)
**Length:** 61.72m/202ft 6in
**Beam:** 7.31m/24ft
**Armament:** 6 x 533mm/21in tubes, 12 x torpedoes; 1 x 76mm/3in gun
**Propulsion:** 2 x 1,156kW/1,550hp diesel, 2 x 969kW/1,300 electric
**Speed:** 13.5 knots (surfaced); 10 knots (dived)
**Complement:** 38 men

LEFT: **"T" for "tremendous". A brilliant class of British boats, the T class came in abundance, saw great war service and continued on for 20 years or more. *Tally Ho*, for example, became the first T-class boat to cross the Atlantic on snorkel.** BELOW: **Profile of a T-class boat of the 1942 vintage.**

# T class

The T class were the first British submarines to have their fuel tanks inside the hull, which eradicated the problems caused by leaking fuel leaving a trail across the sea's surface. Intended to replace the O, P and R classes and 406 tonnes/400 tons smaller, they were superior in nearly every respect to those they were succeeding. They had better armament and two more torpedo tubes, and their all-welded hull meant that the boats were much stronger and able to dive deeper to escape attack and counter-attack. The London Naval Treaty of 1935 had placed restrictions on Britain's submarine ambitions, so these craft had to be constructed within the rules laid down.

LEFT: **Wartime T-class boats carried a 102mm/4in guns, and later in the war, three machine-guns were added.**

There was a slight loss of speed, for example, as the size of the submarine was limited by the treaty, so larger engines could not be fitted. A variety of engines were used to try to improve performance, but with the outbreak of war only Admiralty or Vickers diesels were fitted. One batch of this class even featured 11 torpedo tubes plus a 20mm/0.79in anti-aircraft gun and 7.7mm/0.303 machine-guns.

One of the T class, *Thetis*, was at the centre of one of the most tragic episodes in Royal Navy submarine history when she sank while on trials in Liverpool Bay in 1939 with more than 100 people aboard, including shipyard workers, office staff and others there to celebrate the chance to sail in a submarine. Another T-class boat, *Tetrarch*, attacked a large merchantman in the North Sea, whereupon she was the subject of a very hostile counter-attack by the escorts. Forced to dive deep, she reached 120m/394ft and stayed there for 43 hours.

*Triton* attacked a German convoy taking part in the occupation of Norway, with a salvo of six torpedoes hitting the *Friedenau*, *Wigbert* and *V1507*, killing 900 German troops who were being sent to occupy the country. *Thistle* attacked U-boat *U4* in the North Sea in April 1940 unsuccessfully – but a few days later *U4* caught *Thistle* on the surface and sank her with the loss of all hands. T boats were heavily involved in the Mediterranean and in the Far East, scoring some spectacular results against the Japanese fleet, particularly *Trenchant* which fired a salvo of all eight of her

ABOVE: **A typical T boat patch.** RIGHT: **An impressive whole boat view of HMS *Thorough*, the first Royal Navy submarine to circumnavigate the globe, when she travelled to Australia through the Mediterranean and the Suez Canal and returned via the Panama Canal.**

bow torpedoes at the Japanese heavy cruiser *Ashigara*. Five struck home, blowing away the bows and setting the ship ablaze before she sank.

*Truant* was the only Royal Navy submarine to sink enemy ships in three different theatres – Home, Mediterranean and Far East – sinking almost 81,284 tonnes/80,000 tons of enemy shipping. Before that, she was returning from a refit in America when she spotted a Norwegian ship acting suspiciously. *Truant's* captain discovered that the ship had already been boarded and was under the command of a German prize crew. As well as this, there were captured British survivors from the SS *Braxby* on board. The British crew were taken aboard *Truant*, along with the captain and his wife, the latter of whom became the first woman to sail in a submarine.

In February 1945, *Tantalus* completed the longest wartime patrol, during which she travelled 18,507km/11,500 miles in 55 days. One of the most bizarre stories surrounding a T-class boat came when *Totem* visited Canada and was in turn visited by members of the Cowichan Indian tribe, who presented the boat with a small totem pole. The sailors were told the totem pole had magic properties provided by the carvings of the thunderbird, grizzly bear, killer whale and Fire God, and while the totem remained on board no harm would come to Totem. The boat was transferred to the Israeli Navy and renamed *Dakar* in 1964 and the carved totem placed on display in the Royal Naval Museum in Gosport. In January 1968, *Dakar* was lost in the Mediterranean. In 1957, *Thorough* became the first submarine to circumnavigate the world.

LEFT: ***Thermopylae* was among the last of the T boats to be built, thus benefiting from the extensive modifications that were progressively applied to the class, especially in electronics and crew spaces.**

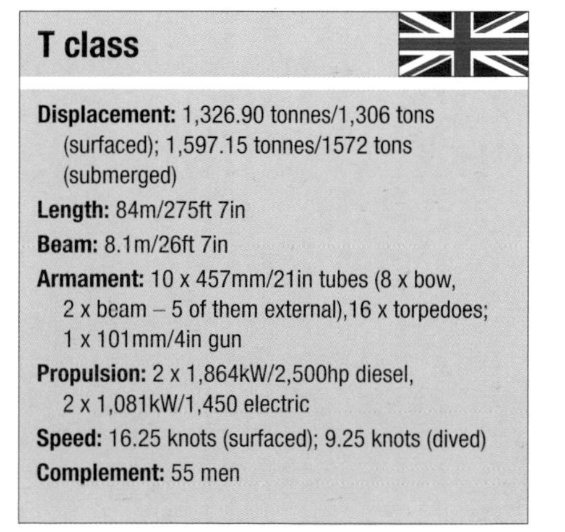

## T class

**Displacement:** 1,326.90 tonnes/1,306 tons (surfaced); 1,597.15 tonnes/1572 tons (submerged)
**Length:** 84m/275ft 7in
**Beam:** 8.1m/26ft 7in
**Armament:** 10 x 457mm/21in tubes (8 x bow, 2 x beam – 5 of them external),16 x torpedoes; 1 x 101mm/4in gun
**Propulsion:** 2 x 1,864kW/2,500hp diesel, 2 x 1,081kW/1,450 electric
**Speed:** 16.25 knots (surfaced); 9.25 knots (dived)
**Complement:** 55 men

FAR LEFT: **The British U-class boats were heroes of World War II. They saw extensive action and became the scourge of the Italian Navy in the Mediterranean.** LEFT: **Naturally, the raising of the Jolly Roger was a common sight at the time, as seen here by *Unbroken*, which torpedoed and damaged the Italian cruisers *Bolzano* and *Muzio Attendolo* on August 13, 1942.** INSET: **A patch for *Unbroken*.**

# U class

Originally designed as unarmed target submarines between the two world wars, the Admiralty quickly changed its mind as alarm bells began to sound across Europe, and it was decided to give them torpedo tubes so they could be used for training crews meant to man the much larger T-class boats. Only three of these boats had been completed by the outbreak of World War II, but their design was so simple it was easy to increase production. Moreover, their small size was considered an advantage for operating in difficult coastal waters, and they proved to be highly successful.

At the outset, the boats had six torpedo tubes with two in an external bulbous bow, but this shape caused wake problems, reduced visibility at periscope depth and was changed by removing the extra two tubes in later models. With fuel and ballast within the pressure hull and diesel-electric power, the shaft was driven at all times by the electric motors, with the diesel engine used on the surface to power the generator that, in turn, topped up the batteries and fed the motor.

The first to be commissioned was *Undine*, which was also the first to be lost when she attacked three German minesweepers in the Heligoland Bight. Her salvo missed, but the torpedo trail gave the enemy a fix on her position, and they counter-attacked. *Undine* dived and, after thinking it safe to return to periscope depth, she did so only to be hit by an explosion that wrecked the hydroplanes. Out of control, the crew had little option but to abandon ship, scuttling the boat with an explosive charge as they left.

U-class boats made a significant contribution to the war effort, particularly *Upholder*, commanded by Lt Cdr Malcolm David Wanklyn, VC, DSO, who sank or damaged 22 enemy ships totalling well over 101,605 tonnes/100,000 tons.

Another remarkable boat, *Unbroken*, used her last four torpedoes to attack an Italian heavy cruiser and a light cruiser which had been bombarding Allied convoys trying to get supplies to Malta. Attacking from close range, she put both ships out of action for the remainder of the war. The escorting Italian destroyers reacted with fury, unleashing a barrage of more than 100 depth charges to try to hit *Upholder*, who survived the onslaught and quietly slipped away to Malta.

*Unbroken* was eventually loaned to the Russian Navy, and several other U-class boats were loaned to other Allied navies. As soon as the war in Europe ended, the U class lost their operational role and reverted to being used as training craft.

LEFT: **The loading of torpedoes and armaments into *Ursula* prior to another patrol, possibly when she hit the Italian liner *Vulcania*, being used for troop transport in September 1941.**

## U class

**Displacement:** 636 tonnes/626 tons (surfaced); 739.65 tonnes/728 tons (submerged)
**Length:** 58.21m/191ft
**Beam:** 4.88m/16ft
**Armament:** 6 x 533mm/21in tubes (bow – 2 x external), 10 torpedoes; 1 x 76mm/3in gun
**Propulsion:** 2 x 298.3kW/400bhp Davy Paxman diesel engines; 2 x 307.2kW/412ehp General Electric motors
**Speed:** 12.5 knots (surfaced); 10 knots (dived)
**Complement:** 30 men

LEFT: **It was a great tragedy that *Porpoise* became the last British submarine to be sunk in hostile action in World War II, after laying mines in the vicinity of Penang, leaving a tell-tale trail of oil sighted by a Japanese spotter plane.**

### Porpoise class

**Displacement:** 1,792.22 tonnes/1,763.9 tons (surfaced); 2,068.58 tonnes/2,035.9 tons (submerged)
**Length:** 87.78m/288ft
**Beam:** 9.12m/29ft 11in
**Armament:** 6 x 533mm/21in tubes, 12 x torpedoes; 50 x mines (although more could be carried in place of torpedoes); 1 x 101mm/4in gun
**Propulsion:** 2 x 1,230.4kW/1,650bhp Admiralty diesel engines; 2 x 607.7kW/815ehp electric motors
**Speed:** 15 knots (surfaced); 8 knots (dived)
**Complement:** 59 men

# Porpoise class

These boats, the Royal Navy's submarine minelayers, went about their business in perilous coastal waters which often left them vulnerable to danger from land, sea, air and their own deadly cargo. Of the six built, four were lost and one captured. Founded on the experience gained with *M3*, the boats featured a chain conveyor and rail system atop the pressurized hull that could carry 50 Mk XVI mines. Because of the extra space intended for mine-carrying, the Porpoise boats were pressed into service to carry vital supplies to the beleaguered island of Malta in 1941–42, with fuel, ammunition and other material occupying the space reserved for the mines.

With the development of a special submarine mine which could be launched using the 533mm/21in torpedo tubes, other Porpoise boats were cancelled. *Rorqual* laid minefields in both the North Sea and Mediterranean, and *Seal*, spotted laying a minefield, was attacked by a German aircraft which also alerted anti-submarine craft. However, before they could reach her *Seal* struck a mine and sank, becoming embedded in the mud on the sea floor. After a struggle she freed herself, but as soon as she returned to the surface, she was attacked, forcing the crew to surrender.

By the end of hostilities, the *Porpoise*-class minelayers had laid more than 2,500 mines, with only *Rorqual* surviving to the end of the war.

LEFT: ***X25*, one of the last of the midget submarines to be built in Britain, after some challenging operations and an exceedingly creditable history.**

### X-craft

**Displacement:** 30.48 tonnes/30 tons (surfaced); 33.02 tonnes/32.5 tons (submerged)
**Length:** 13.26m/43ft 6in
**Beam:** 2.44m/8ft
**Armament:** 2 x containers with 1,620kg/3,571.5lb of explosives; Limpet mines
**Propulsion:** 1 x 31.3kW/42bhp Perkins diesel engine; 1 x 18.6kW/25ehp electric motor
**Speed:** 6.5 knots (surfaced); 4.5 knots (dived)
**Complement:** 3–4 men

# X-craft

Initially, when Winston Churchill was First Lord of the Admiralty, he and the Admiralty looked down their noses at the concept of midget submarines. They were dangerous and the sort of power a weaker nation would employ – certainly not the Royal Navy! However, come World War II and the German threat to both British shipping and the country itself, particularly from the massive German warships operating with seeming impunity from the fjords of occupied Norway, the Admirals began hailing midget submarines as the solution.

Built, developed and trialled under conditions of great secrecy, the midget submarines were crewed by a special breed of men, many of whom were to lose their lives in the service of their country as they struck at German capital ships and installations in seemingly impregnable locations.

After World War II, many films and books were made about the gallant men of the X-craft. Not only did they face the dangers of enemy fire, but they also had to brave minefields, anti-submarine nets, acoustic alarms and other perils just to reach their target before dropping their explosives and trying to find their way out again to rendezvous with their escorts.

For their gallantry in wartime, X-craft crews won 4 Victoria Crosses, 11 DSOs, 17 DSCs, 12 DSMs, 6 CGMs, 4 BEMs, 3 CBEs, 1 OBE, 10 MBEs and 27 Mentions in Despatches.

# A-class SSK

LEFT: **HMS *Artful*, one of the A-class boats that were the backbone of the British submarine fleet in the immediate post-war years.**

Work began in 1942 to produce a boat with superior speed, range and crew conditions, designed especially for use in the Far East. Based on the T class, but incorporating the lessons learned in war conditions, and even including air conditioning, the boats represented a leap forward. With fully welded hulls and a heavy armament of six forward-firing torpedoes and four firing aft, coupled with an air-warning radar which worked while the boat was submerged, they were a vast technological improvement. They also had the "Snort" breathing capability.

The Admiralty ordered 46 A-class boats with the first, *Amphion*, launched in August 1944. *Astute* was next and these were the only two delivered before hostilities ceased. At first they were unstable, but an extra buoyancy tank cured the problem. When Japan surrendered, 30 of the A-class boats were cancelled. But a new type of war – the Cold War – faced the world, and the 16 A-class boats were fully deployed in improving submarine warfare by undergoing strenuous trials and deployments. *Alliance* and *Ambush* were used to carry out exercises in endurance underwater as they stayed submerged for long periods. *Alliance* sailed submerged for 30 days, and in February 1948, *Ambush* sailed from Rothesay on a submerged patrol that was to last six weeks.

Other craft were submerged – unmanned – to see how deep they could go before the hull failed. *Andrew* sailed submerged for 4,023 km/2,500 miles from Bermuda to England in 15 days, the first submerged trans-Atlantic passage. She was also the last Royal Navy submarine to be fitted with a permanent deck gun.

### A-class SSK

**Displacement:** 1,497.16 tonnes/1,473.5 tons (surfaced); 1,645.92 tonnes/1,619.9 tons (submerged)
**Length:** 85.85m/281ft 8in
**Beam:** 6.78m/22ft 3in
**Armament:** 10 x 533mm/21in tubes (6 x bow – 2 x external, 4 x stern – 2 x external), 20 x torpedoes or 26 mines; 1 x 101mm/4in gun, 1 x 20mm/0.79in anti-aircraft gun
**Propulsion:** 2 x 3206.5kW/4,300bhp Admiralty or Vickers diesel engines; 2 x 932.1kW/1,250ehp English Electric electric motors
**Speed:** 18.5 knots (surfaced); 8 knots (dived)
**Complement:** 60 men

LEFT: **HMS *Walrus*, of the Porpoise class, noted for its clean welded hulls and for being fast and silent underwater, undoubtedly among the best conventional boats in the world at the time.**

# Porpoise class (Late)

If the A class was considered a leap forward, the Porpoise was a giant one. As the first post-World War II designed boat to enter service, it was capable of high underwater speed, very deep diving and was very quiet. It was fitted with very large batteries to store the power required, and with its advanced snort and replenishment systems could operate independently without support anywhere in the world for months – exactly what the old time submariners had dreamed of decades earlier.

Porpoise had both air and surface warning radar which could operate at periscope level and while on the surface, as well as an advanced snort system which allowed maximum charging even in rough seas. A bonus was that this provided adequate air conditioning for the boats to operate in Arctic or tropical environments and kept it dry internally.

Periods of six weeks underwater were also made possible because of its oxygen replenishment system and the carbon dioxide eliminators. As well as being equipped to distil fresh water from seawater, they had enough space to carry substantial stores to enable them to operate far from home and far from support for long periods.

In general, they were excellent, seaworthy boats superbly manufactured. However, they were already out of date, at least in terms of Britain's ambitions towards being a nuclear power, and these diesel boats began to take a back seat.

### Porpoise class (Late)

**Displacement:** 2,062.48 tonnes/2,029.9 tons (surfaced); 2,443.48 tonnes/2,404.9 tons (submerged)
**Length:** 89.99m/295ft 3in
**Beam:** 8.08m/26ft 6in
**Armament:** 8 x 533mm/21in tubes (6 bow, 2 stern), 30 x torpedoes
**Propulsion:** 2 x 1,230.4kW/1,650hp Admiralty Standard Range diesel engines; 2 x 3,728.5kW/5,000ehp electric motors
**Speed:** 12 knots (surfaced); 17 knots (dived)
**Complement:** 71 men

LEFT AND BELOW LEFT: **HMS *Explorer* was an experimental boat, but initially had so many problems that her first captain was never able to take her to sea. The issues were eventually resolved, but the boat was scrapped.**

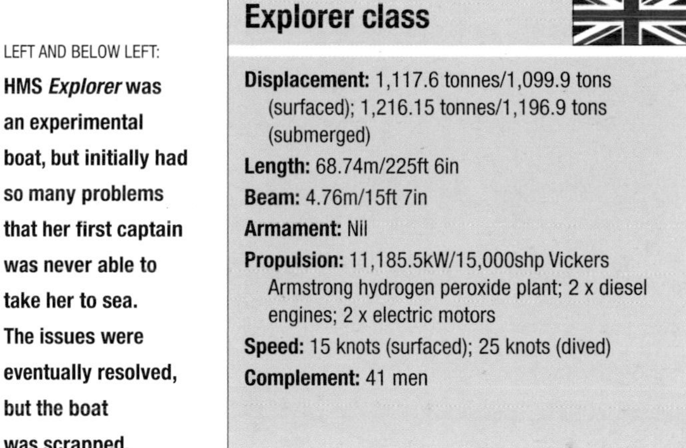

## Explorer class

**Displacement:** 1,117.6 tonnes/1,099.9 tons (surfaced); 1,216.15 tonnes/1,196.9 tons (submerged)
**Length:** 68.74m/225ft 6in
**Beam:** 4.76m/15ft 7in
**Armament:** Nil
**Propulsion:** 11,185.5kW/15,000shp Vickers Armstrong hydrogen peroxide plant; 2 x diesel engines; 2 x electric motors
**Speed:** 15 knots (surfaced); 25 knots (dived)
**Complement:** 41 men

# Explorer class

Designed to trial a new diesel-electric powerplant that incorporated hydrogen peroxide in its fuels, these were believed to be the fastest submarines in the world at the time. The main propulsion was provided by turbines powered with steam and carbon dioxide, which enabled the boat to produce full power while submerged, independent of any external air source. In addition, the *Explorer*-class boats were fitted with conventional diesel engines for use on the surface and battery-powered electric motors for underwater operations.

Only two boats in this class were built. With retractable fittings improving their already streamlined shape, they had excellent manoeuvrability combined with a high submerged speed. However the fuel was highly unstable, and there were several explosions. As the boats were purely experimental, they were unarmed, and their crews were accommodated in converted minesweepers for safety reasons. Capable of 25 knots submerged, they were very fast. But once the Americans succeeded in building a nuclear reactor for a submarine, the project with the Explorers was quickly terminated.

LEFT: **HMS *Orpheus* of the Oberon class was Britain's first major step into modern design, with improved detection equipment and the ability to fire homing torpedoes. Known for their reliability and quietness, many Oberons were sold to overseas buyers.**

## Oberon class

**Displacement:** 2,062.48 tonnes/2,029.9 tons (surfaced); 2,448.56 tonnes/2,409.9 tons (submerged)
**Length:** 89.99m/295ft 3in
**Beam:** 8.08m/26ft 6in
**Armament:** 8 x 533mm/21in tubes, Mk 8 torpedo, Tigerfish torpedo; Sub-Harpoon anti-ship missiles
**Propulsion:** 2 x 2,237.1kW/3,000bhp Admiralty Standard Range diesels; 2 x 1,280kw/1,716.5hp electric motors
**Speed:** 12 knots (surfaced); 17 knots (dived)
**Complement:** 71 men

# Oberon class

Although almost identical to the Porpoise class, the Oberons were fitted with improved detection equipment and were equipped with homing torpedoes. In addition, for the first time in a Royal Navy submarine, plastic was used and glass fibre was incorporated on part of the bridge superstructure and casing. *Oberon*, the first of the class, was launched in 1959. In 1970, during trials in the Mediterranean with a submarine escape-team, a world record was established when the men "escaped" from *Osiris* while she was moving at a depth of 182.9m/600ft.

Only one *Oberon*-class boat – *Onyx* – saw active service when she was the only diesel submarine sent to the Falklands as part of Operation Corporate, and Historic Warships now preserve her at Birkenhead.

In the 1980s, the fitting of sonar and the capability to fire Tigerfish torpedoes and Harpoon anti-ship missiles enhanced the performance of Oberon boats, and the submarines had the ability to guide two torpedoes simultaneously. *Otus* also took part in a series of trials in a Norwegian fjord to determine at which depth submariners could safely expect to escape from a stricken submarine. Two sailors managed to "escape" from 182.9m/600ft yet again. A number of the boats were eventually leased and then sold to Canada.

LEFT: ***U14*** **was commissioned in 1912 with the altered specifications introduced with** ***U13*****, but while serving as part of the I Flotilla, she was damaged by gunfire from the armed trawler** ***Oceanic II*** **and sank off Peterhead, with one member of crew killed and 27 rescued.** BELOW: ***U9*****, noted for reloading torpedoes while submerged, for the first time in history, and later for sinking three British cruisers,** ***Aboukir*****,** ***Hogue*** **and** ***Cressy*** **with an hour.**

# *U1* (Early)

German shipbuilders were producing submarines for the Russian Navy a full two years before the first of what would prove to be a long and successful progression of U-boats made her appearance. *U1* was launched from the Germaniaweft yard in August 1906, and commissioned primarily for trials and evaluation by the Imperial German Navy. Already larger than the British and American counterparts of the A and B classes, *U1* did not, however, impress her own Naval chiefs. They complained that her displacement rendered her unfit for operations any distance from the coast, and modifications began immediately. Even so, *U1* was the precursor to 375 U-boats of three principal classes ranging through 29 different types available for World War I. These new offerings also came with the distinct advantage of a heavy oil motor that precluded the internal problems experienced by petrol-driven engines. However, the downside of this development was that the boat billowed smoke and sparks from a rear-mounted upper-deck exhaust system, emissions that were more on a par with a steamer. Thus, any opposition would have no problem in tracking her on the surface.

This was soon to be modified with a development that dramatically changed surface propulsion forever, with the introduction of diesel engines for the bulk of the U-boats built by Germany. *U1* did impress in other areas, however, and notable among her early trials was a 1,087km/587-nautical-mile endurance test from Wilhelmshaven that took her around Denmark and back to Kiel in exceedingly dire conditions. Based upon these findings, Germany initiated a rapid programme of development and innovation, and by 1911, U-boats had progressed from 42.4m/139ft to 57.3m/188ft in length with subsequent upgrades in both speed, endurance and firepower, increasing her torpedo payload from one to four tubes, two bow and two stern. Further types in this series emerged with additional design improvements through to *U18*.

The fortune of war, however, did not favour these early submarines of the Imperial Navy. Four days after the United Kingdom declared war on the German Empire on August 4, 1914, ten boats, *U5*, *U7*, *U8*, *U9*, *U13*, *U14*, *U15*, *U16*, *U17* and *U18* were despatched to launch an audacious attack on the British Grand Fleet at Scapa Flow. Two boats, *U5* and *U9*, had to return to port with engine trouble, *U13* was lost without trace, and *U15* went down with all hands after being rammed by the British cruiser HMS *Birmingham* in the North Sea on August 9,

LEFT: **Loading torpedoes on the World War I training boat** ***U25*****, which survived the war to be surrendered and broken up at Canning Town in 1922.**

LEFT: **A classic silhouette of *U35*, one of the most successful boats of World War I, which, during 17 patrols with the II Flotilla in August 1915, is credited with sinking 224 ships for a total of 540,402 tonnes/539,741 tons, and survived the conflict, to be surrendered and broken up at Blyth.** ABOVE: **An internal view of the engine room.**

1914. This unpromising start was soon to be repealed. On July 16, 1914, the ageing *U9*, commanded by Otto Weddigen, became famous in the German submarine fraternity by performing a torpedo reload while submerged, for the first time in history. Weddigen began drilling his crews in fast reloads by moving members of his crew forward then aft while the ballast was regulated. On September 22, he used this technique to sink three British cruisers in an hour. It was an unique achievement in that after the first strike, the other British ships sailed to assist the stricken vessel thinking that it had hit a mine, unaware of *U9*'s presence. The succession of strikes occurred at 07:20 hours (*Aboukir*), 07:55 hours (*Hogue*) and 08:20 hours (*Cressy*), each one with a single shot, and claiming a total of 1,400 lives. A few days later, *U9* torpedoed another British cruiser, *Hawke*, and in doing so, this one submarine had killed more British sailors than were lost by Lord Nelson in all his battles put together.

As for *U1*, although no wartime patrols are recorded, she survived the conflict and was decommissioned in the post-war surrender of German U-boats to the Allies. She was thereafter sliced in two and retained for posterity in the *Deutsches Museum* in Munich. The improvements achieved in the designs in the first 18 boats prior to the introduction of diesel propulsion can be seen by comparing the specifications below.

ABOVE: **A U-boat flotilla (left to right): *U13* was lost with all hands in 1914; *U5* survived the war after 22 patrols; *U11* was a training boat; *U3* was a training boat which surrendered in 1918 but sank on the way to the UK; and *U16* sank ten ships during the war, then sank on the way to surrendering.**

## *U1*

**Displacement:** 234 tonnes/230.3 tons (surfaced); 278 tonnes/273.6 tons (submerged)
**Length:** 42.39m/139ft
**Beam:** 3.75m/12ft 4in
**Armament:** 1 x 450mm/17.7in tube (bow), 3 x torpedoes
**Propulsion:** 298.3kW/400hp heavy oil; 298.3kW/400hp electric
**Speed:** 10.8 knots (surfaced); 8.7 knots (dived)
**Complement:** 12 men (22, then 28 on larger types)

## *U18*

**Displacement:** 555 tonnes/546.2 tons (surfaced); 680 tonnes/669.3 tons (submerged)
**Length:** 62.35m/204ft 7in
**Beam:** 6m/19ft 8in
**Armament:** 6 x 450mm/17.7in tubes (2 x bow, 2 x stern); 1 x 10.5cm/4.13in deck gun

RIGHT: **Impressively sleek for the era and suitably menacing, an array of early German submarines, from left to right in the front row, *U22, U20, U19* and *U21* pictured at Keil harbour, part of Germany's High Seas Fleet of 20 boats at the beginning of World War I. They were subsequently deployed as a defensive screen in the North Sea.**
ABOVE: **Profile of *U21*, the first U-boat to sink a British warship, the cruiser HMS *Pathfinder.***

# *U19*: diesel

Innovations in U-boat construction continued at a rapid pace with the first four U-boats designed specifically for the ocean-going hunter role, entering service from 1913 with the leader boat of this type, *U19*.

Her commissioning brought another landmark in submarine development in that she was the first to be manufactured with a diesel engine, so-called after the German engineer who invented it. Even so, by 1914 Germany had only 24 boats in service, four others under repair or seconded to a training role and another 17 under construction, representing one of the smallest fleets among the major powers.

However, it had become clear that coming late to submarine warfare had enabled the Germans to learn from the costly lessons experienced by other navies, at least in design if not immediately in operational tactics. This benefit was beginning to show in the development of their first oceanic boats whose clear aim would be to attack merchant ships in wartime.

New targets had been set for endurance and surface speed, and it became evident, not least in the use and placement of deck guns, that these were to become an important element in surface operations against merchant shipping. Some boats of the new class were equipped with top-mounted guns that folded neatly away after operations, thus allowing greater speed.

It was also a noticeable tactic in the first ship attacks that torpedoes were only used on high profile targets. One such target appeared in the periscope vision of *U21* commander Otto Hersing off the Scottish coast on September 5, 1914. He opened fire to become the first submarine to release a torpedo in anger in World War I. The victim was the British light cruiser, *Pathfinder*.

Hersing went on to become one of the most successful U-boat commanders, sinking or disabling almost 40 ships in 21 patrols over the next three years, a run that included Allied battleships while assigned to assist the Turks during the Gallipoli crisis.

More famously, a sister ship in this class, *U20*, torpedoed the Cunard liner *Lusitania* on May 7, 1915, with the loss of nearly 1,200 passengers and crew.

RIGHT: **Given the prolific success record of *U35*, the submarine was also leading the league table of torpedoes fired. This series of U-boats carried only six torpedoes, so it required a constant re-supply, often at sea.**

### *U19*

**Displacement:** 546.25 tonnes/537.6 tons (surfaced); 669.25 tonnes/658.7 tons (submerged)
**Length:** 62.35m/204ft 7in
**Beam:** 6.10m/20ft
**Armament:** 6 x 457mm/18in tubes; 1x 10.5cm/4.13in deck gun
**Propulsion:** 1,305kW/1,750hp diesel; 894.8kW/1,200hp electric
**Speed:** 15.4 knots (surfaced); 9.5 knots (dived)
**Complement:** 35 men

# U81

By 1914, Germany's U-boat fleet had numerically reached *U38* but as the war progressed, production doubled and then trebled, along with numerous changes in type, layout and variations of armament. However, the key to German submarine expansion was the drive for longer-range boats, introduced in numerous types built in small classes, which consequently allowed the introduction of new features and capabilities. By 1916, the series of 46 Mittel-type boats that began with *U81* was well underway, providing ever-increasing endurance and firepower, which enabled the expansion of operations outward from European waters into the Atlantic and ultimately in threatening posturing towards the east coast of America.

The capabilities of these boats were geared towards long patrols, packed to the gills with torpedoes. The 813-tonne/800-ton U81 series carried ten torpedoes and a substantial deck armoury to a range of up to 12,070km/7,500 miles, while their successors in the U90 and U100 series could take up to 16 torpedoes, although their endurance was significantly lower.

Given the possibilities for causing huge damage to the British, the Germans declared unrestricted warfare on February 29, 1916, under which enemy armed vessels could be attacked anywhere and merchant ships in the war zone could be attacked without warning. This was curtailed only after strong protests from the Americans following the sinking of passenger vessels and the death of American citizens.

The retraction once again forestalled American entry into the war, but it had little effect on the mounting U-boat toll as the German submarine fleet continued to expand both numerically and in capability. *U81*, for example, sank 31 ships – excluding enemy warships – of 91,444 tonnes/90,000 tons on just four patrols, between October 1916 and May 1917, before she herself was sunk by the British boat *E24* off the west coast of Ireland, while *U90* took down 35 ships in the last year of the war.

**ABOVE: An interesting view of *U117*, a minelayer of the UE2 class of which nine boats were commissioned. Although not operational until March 1918, she sank 24 Allied vessels (warships excluded). She was surrendered to the Americans and was used for exhibitions.**
**ABOVE LEFT: A profile of Type U139, a cruiser merchant class of which three were built.**

**ABOVE: The mechanics of the ocean minelayer *U117*, capable of carrying 42 mines, as well as 14 torpedoes, thus providing not only greater firepower but also endurance at sea.**

### U81

**Displacement:** 795 tonnes/782.4 tons (surfaced); 934 tonnes/919.2 tons (submerged)
**Length:** 70.06m/229ft 10in
**Beam:** 6.30m/20ft 8in
**Armament:** 6 x 508mm/20in tubes (4 x bow, 2 x stern), 16 x torpedoes; 1 x 8.8cm/3.46in and 1 x 10.5cm/4.13in deck guns
**Propulsion:** 1,789.7kW/2,400hp diesel; 894.8kW/1,200hp electric
**Speed:** 16.8 knots (surfaced); 9.4 knots (dived)
**Complement:** 39 men

LEFT: **Class leader *UB1*. Built primarily for coastal work, although they did not come into production until the start of the war, they certainly made up for lost time in terms of patrols completed. A total of more than 100 patrols was not uncommon, although the casualty rate was high.** BELOW: **A profile of a UB-class boat.**

# UB classes I, II and III

While remarkable advances were being achieved by Germany in the construction of larger, long-range boats, there was a yawning gap in the coverage of areas where endurance and range were less critical. With a build time of up to 18 months for ocean-going U-boats, in the interests of sheer expediency, a smaller boat was required to undertake missions closer to home. A solution was quickly instigated under which two new classes of smaller submarines were designed, the UB and UC classes, the former to operate principally in the North Sea and coastal regions and the latter designed specifically as minelayers.

The great advantage of this policy was the dramatic cut in the build time. Initially, the UBIs at 125-tonne/123-ton surface displacement were just a quarter of the size of the newest U-boats and could be built in a fifth of the time by having prefabricated sections constructed in Germany and assembled in Belgium. The overall construction time was cut to 80 days for the first of the UB boats, with the first entering

service in the late spring of 1915, thereafter to be built in large numbers through three classes to accommodate ongoing improvements and higher specifications. The fact was the UBI series was underpowered, with a 44.7kW/60hp diesel engine turning a single screw.

They were slow and difficult to handle, and 17 of the UBI range were built before progressive improvements in almost every area were initiated. UBII was better and was extensively used, but still suffered from the limitation of

LEFT: **The familiar cigar shape of *UB64*, commissioned in 1917 in the UBIII series. It operated in the 5th and 2nd Flotilla respectively and claimed 30 ships sunk (excluding warships) during eight patrols in her 13-month lifespan, prior to being surrendered.**

being underpowered. This was evident from the fact that of the 30 built, only seven survived the war. Of those, *UB29* became famous for two reasons: in March 1916, she torpedoed and sank a French cross-channel steamer, *Sussex*. An international furore ensued when it was learned that among the 80 civilians who died were 25 Americans, although even that did not persuade the US to enter the war. In December the same year, *UB29* became the first boat to be sunk by a depth charge during action in the English Channel.

The final model in this class, UBIII, was a larger improvement on the previous two versions in every respect, but given the risk of the heavily mined areas in which they operated, the mortality was only slightly improved. Even so, of the 95 commissioned, 41 were lost or damaged, more than half of which were mined during coastal patrols.

TOP RIGHT: **A member of the UBIII class, possibly *U127*, which later disappeared, probably mined south of Fair Isle, with the loss of all 34 crew.** ABOVE: **The crew of an unidentified UB-class boat pay considerable interest in a passing *Mittel*-class boat.** BOTTOM: **A meeting at sea – *UB11*, a training boat, pulls alongside *U35*.**

## UBI class

**Displacement:** 125 tonnes/123 tons (surfaced); 140 tonnes/137.8 tons (submerged)
**Length:** 28.10m/92ft 2in
**Beam:** 3.15m/10ft 4in
**Armament:** 2 x 508mm/20in tubes (bow), 4 x torpedoes
**Propulsion:** 44.7kW/60hp diesel; 89.4kW/120hp electric
**Speed:** 6.5 knots (surfaced); 5.5 knots (dived)
**Complement:** 14 men

## UBII class

**Displacement:** 258 tonnes/253.9 tons (surfaced); 287 tonnes/282.5 tons (submerged)
**Length:** 36.10m/118ft 6in
**Beam:** 5.80m/19ft
**Armament:** 4 x 508mm/20in tubes (bow); 1 x 8cm/3.15in deck gun
**Propulsion:** 179kW/240hp diesel; 164.1kW/220hp electric
**Speed:** 9.5 knots (surfaced); 5.8 knots (dived)
**Complement:** 23 men

## UBIII class

**Displacement:** 508 tonnes/500 tons (surfaced); 650 tonnes/639.7 tons (submerged)
**Length:** 53.30m/174ft 10in
**Beam:** 5.80m/19ft
**Armament:** 3 x 508mm/20in tubes (2 x bow, 1 x stern), 16 x torpedoes; 1 x 8cm/3.15in deck gun
**Propulsion:** 179kW/240hp diesel; 164.1kW/220hp electric
**Speed:** 13.5 knots (surfaced); 7.8 knots (dived)
**Complement:** 34 men

LEFT: **The *UCI* coastal minelayers, which progressed through three classes, with modifications at each stage and for which 105 boats in total were either commissioned or being built when the war ended. The tally of Allied ships sunk through the mines that they laid was a horrendous total.** ABOVE: **A profile of a UC-class boat.**

# UC class (Minelayers)

Built with similar urgency to the UB boats, this class was a mining specialist destined largely for coastal work. The same problems also occurred with the UC boats in that the first in production were substantially underpowered, with the result that only 15 of the first series, UCI, were completed before UCII boats went into production in June 1915, with an order for round-the-clock manufacture in five shipyards.

A staggering 63 boats of the improved version were commissioned between June 1916 and January 1917 and with hurried training largely carried out on the older boats, the UCIIs took the brunt of the effort in this vital arena. From a German perspective, the UC crews made a courageous contribution.

Minelaying activities around the British Isles and the Mediterranean proved to be one of the most hazardous of all the diverse operations of German submariners. Of the 15 boats in the first class, only one survived to be surrendered in 1918. At least five fell victim to either their own mines or others laid by their own side.

In class II, the toll was equally horrific. Of the 63 UCIIs built, only 15 survived to complete the act of surrender after the Armistice was signed. This fate befell most of the UCIII version, which was introduced with a more powerful motor, but production came too late for the German war effort, and many were still under construction when the war ended.

LEFT: **A mine taken from the coastal minelayer *UC5*, after the boat was grounded on the infamous Shipwash Shoal in the North Sea. She was scuttled, but the charges failed to explode and she fell into Allied hands.**

### UCII

**Displacement:** 410 tonnes/403.5 tons (surfaced); 485 tonnes/477.3 tons (submerged)
**Length:** 39.47m/129ft 6in
**Beam:** 5.40m/17ft 9in
**Armament:** 3 x 508mm/20in tubes (2 x bow, 1 x stern), 16 x torpedoes; 1 x 8.8cm/3.46in deck gun; 18 x mines
**Propulsion:** 372.8kW/500hp diesel; 313.2kW/420hp electric
**Speed:** 11.6 knots (surfaced); 7 knots (dived)
**Complement:** 14 men

# U151 class

The story of the appearance of a septet of giant and lethal U-cruisers that began to cause havoc among Allied shipping in late 1917 had its beginnings in what was a strangely peaceful mode in June 1916. It was in that month that Britain, and the Americans for that matter, were surprised by the arrival on the other side of the Atlantic of what was heralded as a civilian submarine, a 1,524-tonne/1,500-ton unarmed cargo-carrying boat christened *Deutschland*. The idea was that huge boats would transport cargoes across to the United States and return with much needed raw materials that were in short supply as a result of the British blockade of Germany.

The Americans were happy to accept such arrangements, and *Deutschland* completed the crossing safely and steamed untroubled into Baltimore, Maryland, in July, and unloaded her cargo. She returned unhindered in August, packed with much-needed supplies.

Encouraged, the Germans sent her on a second trip to New London, Connecticut. However, this voyage was less profitable, as the Americans were beginning to get agitated over Germany's application of unrestricted warfare. Consequently, the cargo carrying mode was abandoned, and *Deutschland* and the rest of her class (*U151–U157*) were promptly converted into attack boats, each manned by 56 submariners and a prize crew of 21.

Outfitted during the summer of 1917, they were rushed into service to become much-feared raiders, literally cruising around the Atlantic for several months at a time. They each carried a relatively light load of 16 torpedoes, given the time they were on patrol, and relied heavily on deck artillery – 15cm/5.9in and 8.8cm/3.46in guns – for surface attacks. They were pointedly put under the command of some of Germany's most experienced and famous U-boat skippers.

Even so, these giants of the deep were difficult to manoeuvre and a nightmare for crews during heavy weather, when they rolled and tossed in a gruesome, uncomfortable manner. In spite of this, the boats scored considerable successes. *Deutschland*, recommissioned as *U155*, stayed out for 105 days on her first patrol, covering almost 16,093km/10,000 miles, sank 19 ships and shelled targets ashore on the Azores. *U151* was notably active along the coastline of America during an eventful 13 week cruise, laying mines and sinking 23 ships. Overall, the six cruisers sank 174 ships for a total of 366,793 tonnes/361,000 tons.

**TOP: *U151* was the lead boat in a class of armed merchant cruisers that bore her name, of which seven – including the famous *Deutschland* – were designed and built as large commercial submarines carrying materials to and from locations otherwise denied to German surface ships because of Allied submarine operations. All seven boats were converted to cruisers after the United States entered the war in 1917. ABOVE: A glimpse inside *Deutschland*, or *U155*, as she later became known.**

**LEFT: The engine room of one of the U-cruisers/merchant boats built early in 1916 and which created much interest when surrendered to the Allies at the end of the war.**

## U151 class

**Displacement:** 1,488 tonnes/1,464.5 tons (surfaced); 1,845 tonnes/1,815.9 tons (submerged)
**Length:** 64.9m/212ft 11in
**Beam:** 8.93m/29ft 4in
**Armament:** 2 x 508mm/20in tubes (bow); 20 x torpedoes; 2 x 15cm/5.9in guns
**Propulsion:** 2 x 298.3kW/400hp diesel engines; 298.3kW/400hp electric motors
**Speed:** 12.4 knots (surfaced); 5.2 knots (dived)
**Complement:** 56 men, plus 21 prize crew

LEFT: ***U1*, pristine leader of a brand new breed of U-boats built under the Nazi emblem and commencing with the IIA class. Commissioned in 1935, she lasted only six months into the war before she was lost with all 24 hands, probably to a mine in the North Sea.** BELOW: **A bird's-eye view of the tower for surface operations.**

# U classes (1939) Types IIA, B, C and D

The Treaty of Versailles at the end of World War I saw the U-boat force decimated as the Allied nations took possession of the best of the surrendered submarines to plunder the technology, and destroyed the rest. Germany was also barred from building or possessing any submarines in the tiny navy she was allowed to retain. Even so, the vast storehouse of scientific and engineering acquired during those years could not simply be wiped away. Indeed, when the time came, it would form the backbone of the resurrection of what would become the greatest submarine force the world had seen, bar none.

In the meantime, German submarine crews secretly trained in Russia and Spain and within their own country under the guise of anti-submarine warfare exercises, which were allowed. Nor was Germany prevented from continuing research into submarine development, and by the early 1930s, observers were reporting that a considerable team of scientists and engineers were employed on numerous projects that would prepare the way for the revival of the nation's submarine fleet. Indeed, it was a vital aspect of Hitler's plans when he announced that Germany would no longer be bound by the Versailles Treaty and began to re-arm.

With no navy to speak of, submarines immediately became a crucial element in the Nazi re-armament, and all the pre-planning that had taken place in earlier years now blossomed into reality at an incredible pace. Shipyards across the nation swung into operation, building rapidly during 1934–35, with the first U-boat commissioned on June 29, 1935. This was the date that marked the arrival of the new *U1*, the lead boat of the

Type IIA, whose eventful fate was to be lost with all hands, probably due to hitting a mine, in the North Sea on April 6, 1940. Five more boats came in the first wave of the new U-boats, before they were upgraded in size through three additional types, with 20 boats in IIB, 8 in IIC and 16 of IID. In the final reckoning, quite a number of this type survived World War II; however, the fate of a greater number was settled by mines and British ships in the North Sea.

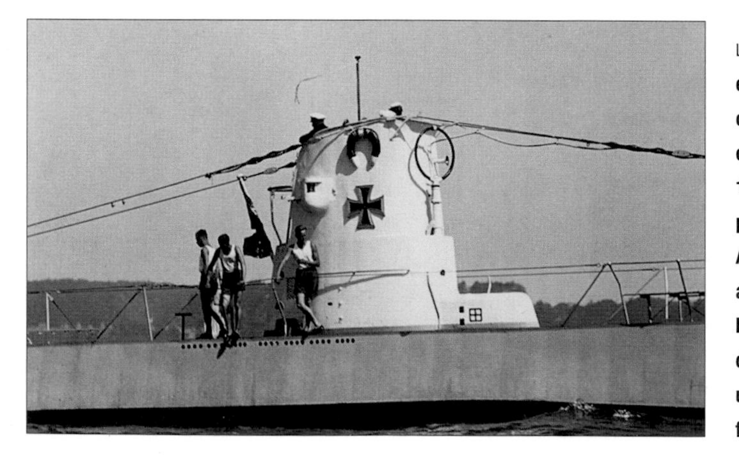

LEFT: **The distinctive emblem, displayed here on the tower of *U9*, originally lead boat of the 1st Flotilla, from which position she sank seven Allied merchantmen and two warships before she herself went down in the Black Sea under a hail of bombs from a Soviet aircraft.**

## U1, Type IIA

**Displacement:** 240 tonnes/236.2 tons (surfaced); 374.9 tonnes/369 tons (submerged)
**Length:** 40.90m/134ft 2in
**Beam:** 3.83m/12ft 7in
**Armament:** 3 x tubes (2 x bow, 1 x stern), 5 x torpedoes
**Propulsion:** 522kW/700hp diesel; 233.7kW/300hp electric
**Speed:** 13 knots (surfaced); 6.9 knots (dived)
**Complement:** 22–24 men

# U-VIIA, B and C

This is the boat that through its three versions became the principal enemy of the Allies for much of the war, especially when operating in wolf-packs. The ocean-going 670-tonne/659-ton Type VIIA, which arrived with *U27* in 1936, was developed from the Finnish Vetehinen design. An improved version came on stream in 1939. These were very efficient machines, agile and popular with their crews. The second version could carry 11 torpedoes or 22 mines, as well as a very effective 8.8cm/3.46in deck gun.

By 1940, the Type VIIC, which became one of the great assets of the German fleet, had overtaken this with more than 600 being built from 1941, beginning with *U69*. Her range was limited to around 10,461km/6,500 miles, but the collapse of France in June 1940 gave the Nazis access to the Atlantic ports, thus providing a haven for 12 U-boat flotillas subsequently to be based in Brest, La Rochelle, La Pallice, St Nazaire, Lorient and Bordeaux, providing access to British waters and the open Atlantic.

A number of the final batch of the VIICs were fitted with the snorkel and another, *U96*, became famous as the U-boat featured in the film *Das Boot*, although the boat was actually sunk by American bombers in Wilhelmshaven in 1945.

The material for the movie resulted from an excursion by Lothar-Günther Buchheim, who was ordered aboard as a war correspondent and artist for propaganda purposes. He took more than 5,000 photographs that survived the war, and his experiences resulted in a best-selling novel.

TOP: **One of Germany's VIIA boats, a new fast attack boat design of which ten were built in the late 1930s. Further improvements in range and firepower were added to the Type VIIB and Type VIIC.** ABOVE: ***U251*, one of the Type VIIC boats that were produced in great numbers from 1941 and became the workhorses of the German U-boat fleet.** BELOW LEFT: **The Type VIIB was given greater fuel-carrying capacity by the addition of external saddle tanks, thus providing a considerable increase in range.**

## U-VIIC

**Displacement:** 755 tonnes/743.1 tons (surfaced); 857 tonnes/843.5 tons (submerged)
**Length:** 67.10m/220ft 2in
**Beam:** 6.20m/20ft 4in
**Armament:** 5 x 533mm/21in tubes (4 x bow, 1 x stern), 14 x torpedoes or 26 mines; 8.8cm/3.46in deck gun
**Propulsion:** 2,386.2kW/3,200hp diesel; 559.3kW/750hp electric
**Speed:** 17.8 knots (surfaced); 7.5 knots (dived)
**Complement:** 44–50 men

# U-IX

Another major development emanating from the 1935–36 period was the introduction of Type IX, an ocean-going submarine with a range of 19,312km/12,000 miles that could travel, when the time came, well into American waters. With 23 torpedoes on board, these boats carried formidable firepower. They were produced in three versions, and the second subtype, IXB, became the most successful type in U-boat history, with some boats accounting for more than 101,605 tonnes/100,000 tons of shipping. The total losses attributed to IXB amounted to 282 ships, displacing 1,551,006 tonnes/1,526,510 tons.

It was also an IXB boat, *U123*, that led the attack in US waters in early 1942 known as Operation Drumbeat, in which five boats sank nine ships in raids along the east coast of America. The marauding boats remained in the area for almost a month and were kept supplied with fuel by a U-boat tanker circling off the coast near Bermuda, an innovation that in this instance extended their patrol for an incredible eight weeks.

But even as these huge campaigns and wolf-pack operations reached their peak in 1943 – a period known among the U-boat men as "The Happy Time" – new scientific developments and sophisticated anti-submarine tactics by the British were already beginning to deliver the required results. This was especially effective in curtailing the operations of IX types, because they took far longer to replace. With mortality rates rising among U-boats of all types, production had to be stepped up to almost impossible levels to keep pace with losses.

Two areas where the British were scoring the most significant results were in the greater protection of convoys, which involved not only surface ships and submarines, but also aircraft equipped with the latest tracking devices. One of the most dramatic examples occurred in May 1943, when a wolf-pack of 12 U-boats attacked a convoy and lost three quarters of their number. The realization began to dawn that the days of such stunning victories against the convoys were coming to an end and, at the very least, would require a significant overhaul of tactics.

One of the Kriegsmarine's attempts to resolve the problem was the order that the IX boats should be fitted with more anti-aircraft guns to attack incoming aircraft, but the RAF simply introduced their own avoidance tactics of staying out of range until the U-boat was about to dive, thus stowing the guns, and then came in for the attack. The switch in operational tactics by the British brought some stunning results, notably in April and July 1943, when 109 U-boats were lost. Coinciding with declining success at sea, U-boat shipyards and pens were now being successfully and regularly bombed by the Allies. It was a major turning point in the fortunes of the German U-boat campaign, and one that resulted in Hitler's intervention, demanding new and better boats and defences. After all, according to his theory, the war should already have been won. As it was, only one IX boat survived the war.

ABOVE: **Under attack: a Type IX, designed as a long-range ocean-going boat, with five external torpedo containers and space for ten additional torpedoes. As mine-layers, they could carry 44 TMA or 66 TMB mines.** INSET: **A profile of *U123*, Type IXB, the most successful of the IX types, with each boat averaging over 101,605 tonnes/100,000 tons of shipping sunk.**

LEFT: **To counter the Allied radar threat, the Germans perfected a snorkel, enabling a submarine to run its diesel engines and recharge its batteries while operating just below the surface.**

### U-IX

**Displacement:** 1,034.40 tonnes/1,018.1 tons (surfaced); 1,160 tonnes/1,141.7 tons (submerged)
**Length:** 76.50m/251ft
**Beam:** 6.70m/22ft
**Armament:** 6 x 533mm/21in tubes (4 x bow, 2 x stern), 22 x torpedoes or 44 mines; 10.5cm/4.13in (45 calibres) deck gun
**Propulsion:** 3,281.1kW/4,400hp diesel; 745.7kW/1,000hp electric
**Speed:** 18.5 knots (surfaced); 7.5 knots (dived)
**Complement:** 48–56 men

LEFT: ***U441*, formerly a VIIC boat rebuilt as *U-Flak1* in 1943, the first of three U-Flak boats, was instantly recognizable by the much larger bridge which accommodated an additional gun platform ahead of the conning tower.** BELOW: **The Flak boat concept was deemed a failure, and instead the VIIC boats were fitted with new anti-aircraft guns.**

# U-Flak boats

This most deadly of contraptions emerged in the spring of 1943, when Germany realized that losses of their submarines in the fiery hell of the Bay of Biscay to Allied aircraft was reaching hugely damaging proportions. The boats were being picked off at an alarming rate as they travelled in and out of the U-boat bases of St. Nazaire, Lorient, Brest and Bordeaux. The region had become commonly known as the Valley of Death among U-boat submariners, and the statistics pinpointing their losses in that area show why they were getting exceedingly worried. More than ten per cent of the entire U-boat losses, or 28 boats, had been sustained in the Bay of Biscay in one year alone, and by the end of the war that total had risen to 72.

After much discussion, the Kriegsmarine came up with what they hoped would be a solution, converting a number of their superb VIIC boats into what amounted to floating flak platforms that could deliver massive firepower against incoming aircraft. Seven boats in all were earmarked for conversion, although only three were completed before the project was called off. They were fitted with two rapid-fire four-barrelled cannon, one placed forward of the conning tower with a shield to protect the gun crew and another located behind the bridge. A further upgrade was to add two heavy machine guns mounted on the bridge. With the additional personnel required for manning the guns and as ammunition handlers, the crew was increased in number from 48 to 69.

Towards the end of May, *U441*, operating under her new name of *U-Flak1*, moved out of Brest to take up her position. Within two days of operations, however, the British realized what was happening and an RAF Sunderland flying boat came over and sent a volley of depth charges that badly damaged the flak boat. Unfortunately, the Sunderland also crashed moments later. *U-Flak1* made an undignified retreat to base but, undaunted, she was hastily repaired and returned to her station in July only to come under almost immediate attack in a pincer movement by three Beaumont aircraft of 248 Squadron, RAF. The boat was again heavily blasted topside, killing ten crew and wounding another 13. The commander had to risk diving with damage to escape total obliteration.

She made it safely back to Brest, but the Kriegsmarine did not give up and ordered three more of the converted flak boats to take up position as the losses in the bay continued to mount. By the middle of August, *U256*, *U271* and *U953* were sent out, but only completed one more patrol before the whole flak-boat concept was abandoned. Instead, the VIIC submarines were fitted with new anti-aircraft guns, and all the flak boats were restored to their former state.

RIGHT: **A U-boat under attack from the air, ever more common as the RAF extended its anti-submarine operations.**

## U-Flak1

**Displacement:** 755 tonnes/743 tons (surfaced); 857 tonnes/843.5 tons (submerged)
**Length:** 67.10m/220ft 2in
**Beam:** 6.20m/20ft 4in
**Armament:** 2 x 2.2cm/0.87in 38/43U four-barrelled cannon; 1 x 3.7cm/1.46in automatic-firing cannon, 2 x MG 42 machine-guns, 8.8cm/3.46in rockets
**Propulsion:** 2,386.3kW/3,200hp diesel; 559.3kW/750hp electric
**Speed:** 17.8 knots (surfaced); 7.5 knots (dived)
**Complement:** 69 men

LEFT: **A sad end for a brilliant boat which was a generation ahead of its time. This is one of the many scuttled or sunk at the time of Germany's surrender and is apparently being recovered. In the years ahead, they were taken apart by the Allies, and the technology purloined.**

# XXI

Ironically, Germany already possessed and was developing technology that might well have won them the war at sea, and certainly the Battle of the Atlantic had it been introduced earlier. New engines were being built, such as the Walther propulsion system that attracted so much attention among the Allies. However, the new star of the show would undoubtedly have been the brand new XXI, a deep-diving combat submarine that had enormous potential, as demonstrated after the war when it became the model for the first nuclear boats. The principal advantages were the super-streamlining of the hull, the snorkel and the huge battery capacity for underwater speed and endurance.

In fact some of the technology had been available since the mid-1930s. The Dutch had already produced snorkel-equipped boats, which fell into both British and German hands after the Blitzkrieg. At the time, neither navy saw any value in them. The thinking among Germany's submarine hierarchy was that U-boats performed better and were safer on the surface after dark, and in the first few years of the war this belief was confirmed. As noted above, the greater use of aircraft combined with radar, in which the British were well ahead in anti-submarine warfare, totally altered this conception. There is no doubt the Germans underestimated the British ASW capability, and in the event, daytime surface running became suicidal.

The solution might have been resolved by Dr Helmut Walther's propulsion system that would create the "true submarine" capable using hydrogen peroxide to power a boat that could operate submerged for a greater period of time, but the project had been shelved through cost and shortage of materials. However, he did resurrect the snorkel idea for modern application, and boats were converted in 1943–44. This permitted them to operate slightly submerged to take in fresh air for the diesels and remove engine exhaust. From there, the snorkel was to be used for a totally new boat, the XXI, with extra batteries in place of Walther's original idea of hydrogen peroxide. Now the Germans had the technology to meet the British counter-measures. It would arrive in the shape of the revolutionary new boat, the XXI ocean-going submarine, and the smaller XXIII coastal version.

The XXI would have electric power three times greater than the VIIC, the recharging process was also considerably shorter, and safer, and a new hydraulic torpedo reload system meant that six tubes could be loaded in the time it took to reload one in the VIIC. Furthermore, the facilities for the crew – always a bugbear among submariners – were excellent, with modern amenities such as individual showers and freezers for food storage.

LEFT: **The sleek lines of the tower of an XXI submarine form the background of this scene of formal welcome for a VIP.**

With its streamlined hull and exciting new features for greater submerged speed and overall efficiency, the boat was heralded as the most important development in submarine warfare, as indeed it was. In 1943 when the tide was definitely turning against the U-boats, Hitler demanded the immediate introduction of the XXI. More than 100 were ordered along with 30 XXIIIs.

Dönitz promised the first would be commissioned within six months. It didn't happen, on account of the increased Allied bombing raids over the shipyards, coupled with chronic shortages of raw materials. The Führer's anger was only placated when his trusted architect Albert Speer agreed to take personal control of the project, but by then the Allies were closing in on all fronts and Germany's war at sea was almost played out. Even so, work on the XXI continued to the last. The first one to undergo trials, *U2511*, went to sea just one week before Germany surrendered and performed in an excellent manner. Others lay in the shipyards, ready and waiting for the Allies to share the spoils – and to contemplate their good fortune that the XXI did not arrive earlier.

TOP: **Another boat before its time, a Type XVII with the Walther turbine. This was an experimental research boat, one of which was transported to the United States, the technology dissected and the boat then used for trials by the US Navy.** ABOVE: **A profile of XXI.** BELOW: **A profile of the Type XXIII, *U2326*, which subsequently became the British submarine *N35*. She sank after a collision in Toulon harbour in 1946.**

ABOVE: **Commissioned as *U2367* in March 1945, this Type XXIII sank before seeing service, after a collision near Schleimünde. She was raised in 1956 and incorporated into the German Federal Navy under the name of *U-Hecht* with pennant number S171, and remained in service for a further 20 years.**

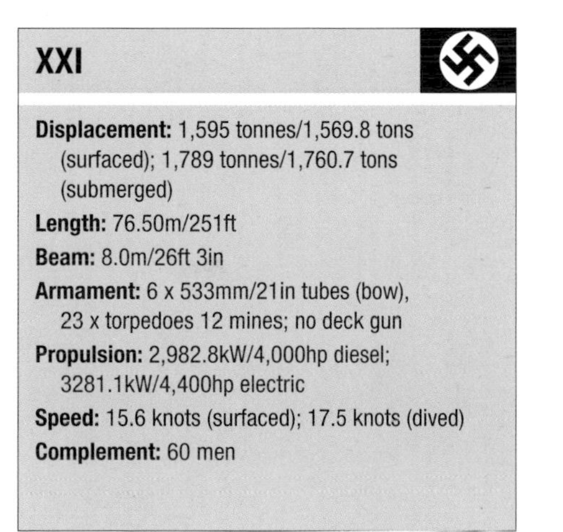

**XXI**

**Displacement:** 1,595 tonnes/1,569.8 tons (surfaced); 1,789 tonnes/1,760.7 tons (submerged)
**Length:** 76.50m/251ft
**Beam:** 8.0m/26ft 3in
**Armament:** 6 x 533mm/21in tubes (bow), 23 x torpedoes 12 mines; no deck gun
**Propulsion:** 2,982.8kW/4,000hp diesel; 3281.1kW/4,400hp electric
**Speed:** 15.6 knots (surfaced); 17.5 knots (dived)
**Complement:** 60 men

# KD classes

The Japanese story in terms of submarines prior to 1945 is one of superlatives and extremes, and thus represents a fascinating array of technology. They produced some of the fastest, largest, longest, widest, heaviest and smallest boats, as well as excelling in aircraft-carrying, long-distance and huge armament, often not exceeded until the arrival of the post-war nuclear fleets. In so many areas their fleets, geared to the vastness of the Pacific Ocean, appeared to have the advantage over any of the other major powers.

Yet, for all that, in the final analysis the Japanese submarine campaign of World War II was widely regarded as an abject failure that even some of their most senior commanders recognized as such. Given the diversity of their fleet and the

ABOVE: **The Japanese Type KD1, built in the inter-war years and based on the surrendered German *U139*, it was used as the prototype for larger I boats (KD2–KD7 types) that came in its wake. A large double-hulled submarine, it was intended to operate with the battle fleet.** BELOW LEFT: **The torpedo tubes of the giant KD3A class.**

spectacular opening of their campaign with the raid on Pearl Harbor supported by 30 submarines, intelligence analysts in the West feared the worst as the Nippon shipyards powered ahead with everything from midget to multi-bomber-carrying boats, and from medium attack boats to giant cruisers. There was also a heavy concentration on endurance, enabling more than 60 of their ocean-going types to travel up to 32,187km/ 20,000 miles or 100 days at sea – a statistic unmatched by any other submarine fleet.

Another remarkable achievement was that of the 56 submarines exceeding 3,048 tonnes/3,000 tons operating in World War II, only four were not Japanese. Yet, tactically they often mystified their opponents, and appeared to be acting in a reconnaissance role to find Allied naval task forces in preparation for the big battles, rather than using their firepower in the hit-and-run role or as wolf-packs, which the British and the Germans in particular mastered so effectively with much smaller boats. Size did matter, and the sheer bulk of these giant craft often proved to be their downfall. They were slow to manoeuvre, sluggish to dive and thus became far easier targets.

The final score card told its own story and indicated the role to which their submarines at large were directed. Japanese submarines were credited with sinking a mere 184 merchant ships throughout the war, compared with the 2,800 claimed by the U-boats. Their losses meanwhile were catastrophic, losing almost three-quarters of her entire fleet, a figure matched only by the U-boats as both those nations faced an onslaught resulting from developments in anti-submarine warfare by the British and Americans. It was also a matter of some satisfaction to the Americans that none of the 30 Japanese submarines identified as having participated in the attack on Pearl Harbor actually survived the war.

The KD classes were among the workhorses of the fleet and had an interesting birth – with a prototype built in 1925, based upon *U139*, which came into the possession of the Japanese after World War I. The KD type progressed through various modifications to the KD7 class over a span of 16 years to 1942. By then a milestone was achieved when the KD6 class produced boats capable of the fastest surface speed of any submarine at the time.

The KD type was moderately successful although there was an early loss of *I70*, sunk by USS *Enterprise* off Oahu on December 10, 1941, in the American strike-back following Pearl Harbor. Overall, the KD classes suffered heavy casualties, with the KD7 class losing all ten boats within a year of coming into service.

TOP: ***I176*** **was one of ten Japanese Type KD7 boats produced by 1942 as attack submarines. They had an outstanding endurance of 75 days and differed significantly from earlier types in that all tubes were sited forward. All were lost by 1944.** ABOVE: **The Type KD4 which were actually smaller than their predecessors of the KD3 class but were said to be unusually agile for Japanese boats and well liked by their crews.** BELOW: **The stylish KD6A, with the highest surface speed of any submarines upon their arrival in the 1930s, although this record was beaten by later Japanese boats. In spite of their speed, the whole class was wiped out by enemy action.**

LEFT: **The KD3A were massive submarines whose design was later revised to enable them to carry a second aircraft. The range of these boats was incredible, but because they were so large, they were easily spotted, slow to dive, easy to track and easy to hit. Consequently losses were also extraordinarily high.**

## KD7 class

**Displacement:** 1,804 tonnes/1,775.5 tons (surfaced); 2,560.88 tonnes/2,530.4 tons (submerged)
**Length:** 105.5m/346ft 2in
**Beam:** 4.57m/15ft
**Armament:** 6 x 533mm/21in tubes (bow), 12 x torpedoes
**Propulsion:** 2 x diesel engines, 5,965.6kW/8,000hp; 1,342.3kW/1,800hp electric
**Speed:** 23 knots (surfaced); 8 knots (dived)
**Complement:** 86 men

LEFT: **Another giant, *I14* was an A-class Modified, the latter to take two Aichi M6A1 seaplanes able to carry a torpedo or an 800kg/1,764lb bomb. Again, the boat had an extensive range but was let down by an indifferent performance underwater.**

# Type A

Only three boats in this class were built, however they are very worthy of mention as a further development of the KD design, but once again much larger even than their forebear, *J3*. The A boats were among the largest built during the war, weighing almost 3,048 tonnes/3,000 tons surfaced and over 4,064 tonnes/4,000 tons submerged.

They were intended as flagship boats with communications systems that allowed them to have command and control of a submarine group, a more formal arrangement of the wolf-pack system.

The A1 design was also used to carry an aircraft, with the hangar pod opening forward of the conning tower, and for easy loading on to the catapult that would shoot the seaplane into the air while at the same time taking advantage of the forward movement of the ship at the point that the aircraft was launched. This development of submarines as aircraft carriers continued through the war, with some spectacular results.

A variation followed with the introduction of the A2 type in 1944 and two substantially larger Modified A2s that carried two seaplane bombers. Once again, however, a complicated diving routine and inert manoeuvrability made these boats an easy target, and none of the class survived more than a couple of years.

**A1**

**Displacement:** 2,872.9 tonnes/2,827.5 tons (surfaced); 4,083.5 tonnes/4,019 tons (submerged)
**Length:** 113.62m/372ft 9in
**Beam:** 9.54m/31ft 4in
**Armament:** 6 x 533mm/21in tubes (bow), 18 x torpedoes; 1 x 14cm/5.5in (50 calibre) deck gun; 1 x seaplane
**Propulsion:** 2 x diesel engines, 9,246.7kW/12,400hp; 1,789.7kW/2,400hp electric
**Speed:** 23.5 knots (surfaced); 8 knots (dived)
**Complement:** 114 men

LEFT: ***I30*, one of the 20 boats of the B class, the largest and busiest group of the Japanese submarine force. Although they were efficient machines, only one survived the war. They were fast and had a forward catapult launch arrangement for their seaplane for a rapid take-off. *I30* was an early victim of Allied mines off Singapore in October 1942.**

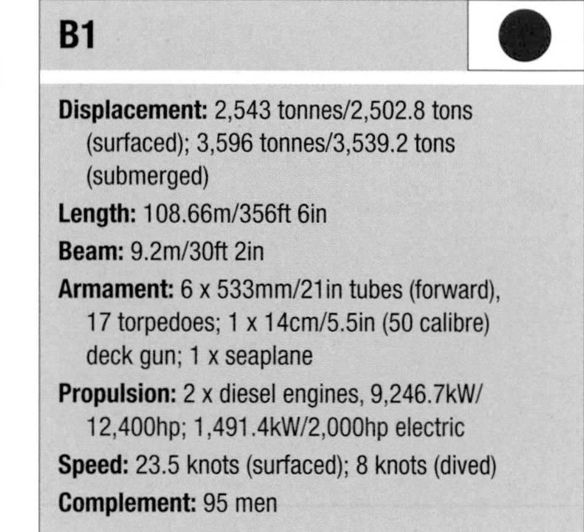

**B1**

**Displacement:** 2,543 tonnes/2,502.8 tons (surfaced); 3,596 tonnes/3,539.2 tons (submerged)
**Length:** 108.66m/356ft 6in
**Beam:** 9.2m/30ft 2in
**Armament:** 6 x 533mm/21in tubes (forward), 17 torpedoes; 1 x 14cm/5.5in (50 calibre) deck gun; 1 x seaplane
**Propulsion:** 2 x diesel engines, 9,246.7kW/12,400hp; 1,491.4kW/2,000hp electric
**Speed:** 23.5 knots (surfaced); 8 knots (dived)
**Complement:** 95 men

# Type B

A busy general-purpose group and numerically one of the largest of the Japanese classes, the B1, B2 and B3 classes were also among the most successful, sinking almost a third of all the merchant ships claimed by Japan during the war, as well as a lion's share of warships. However, in time, they also suffered badly, with a 95 per cent loss rate.

Not quite a monster of the deep compared with others built around the same time, these boats were relatively fast, quicker to react and capable of an impressive range of around 24,140km/15,000 miles at 16 knots. They also had a seaplane tucked away for catapult launching, but on this class the aircraft was aft-mounted. Later in the war, some of the aircraft were removed to provide space for more effective deck armaments, and in early 1945 two of the B3s – *I56* and *I58* – were converted to carry four Kaiten human torpedoes for special operations.

The B boats recorded a number of high profile successes. Of particular note was an action by *I19* in September 1942 when a salvo of six torpedoes sank two American warships and damaged another. She fired the salvo at the aircraft carrier USS *Wasp*. Two struck the giant ship forward and hit the fuel storage tanks, which exploded upon impact. The rest of the torpedoes sallied forth for a further 3,200m/3,500yd into the midst of another US carrier group, this time sinking a destroyer with one shot and causing sufficient damage to the battleship *North Carolina* for her to be withdrawn for major repairs.

The most infamous of the B boat attacks, however, came in early November 1942, when a number of US warships were limping away from the first naval battle of the Guadalcanal campaign. Among them was the light cruiser USS *Juneau* that had been torpedoed on her port side near the forward fire room. The shock wave from the explosion buckled the deck, shattered the fire control computers, and knocked out power. On November 12, however, the wounded American ships were being shadowed by the B type submarine *I26*. At 11:01 hours, the submarine fired a three-torpedo salvo at the cruiser *San Francisco*. None hit that target, but one passed beyond and struck *Juneau* close to her previous damage. The magazine exploded and blew the cruiser in two. She sank within minutes, although 115 of *Juneau*'s crew survived the explosion. However, because of uncertainty about the number of Japanese ships in the area, rescue efforts did not begin for several days.

Exposure, exhaustion and shark attacks whittled down the survivors, and only ten men were subsequently rescued from the water eight days after the sinking. The casualties included five members of one family, now written into US naval history as the Tragedy of the Sullivan Brothers. Many memorials have since honoured the brothers and caused the US Navy to review its policy of allowing family members to serve together at sea.

Thereafter, the fortunes of the B boats began to fade as the US Navy hunted them down. Between August and November 1943, 15 were sunk and others went in early 1944. By the end of the war, there was only one survivor of the B-type boats.

# Type C

The Japanese built a large number of midget submarines between 1936–38, and the C boats, originating from the Type KD6 design, were among the classes capable of carrying them. They were active in that role from the outset, delivering five midgets to participate in the attack on Pearl Harbor. Only one of the midget-carrying boats actually entered the harbour, and that was promptly sunk after she had fired her two torpedoes, which missed their target.

ABOVE: ***I16*** **was one the key players in the transportation of midget craft into attack zones, including Pearl Harbor, and was among the most successful of all the large Japanese submarines in conventional battle. With *I19*, she sank the carrier USS *Wasp* and the destroyer *O'Brien*. *I16* was herself lost in 1944.**

Numerous attacks followed, and in many instances the success rate was less than encouraging for the Japanese Navy. At the end of May 1942, for example, *I16*, *I18* and *I20* spearheaded an attack in which midgets were unloaded for an attack on a British base at Diego Suarez, Madagascar. Only two of the three midgets made it to the attack zone, and torpedoes from one of them hit both of their targets, damaging the battleship HMS *Ramillies* and sinking a tanker. However, in the return fire, both midgets were sunk, and the crews were killed as they attempted to escape to shore. Later that month, five midgets were launched in an attack on Sydney Harbour but failed to score a hit, and all were lost. A similar result came when using midgets despatched from *I16*, *I20*, and *I24* on November 7, 1942, at Guadalcanal. The net result was all five midgets lost in return for damage to one US destroyer.

None of the midget-carrying C1 Type boats survived the war, most falling victim to US warships.

ABOVE: **HMS *Ramillies* was damaged by torpedoes from midget submarines carried by the Japanese submarine *I16* during the bombardment by British warships of Diego Suarez, Madagascar, in May 1942.**

## C1 Type

**Displacement:** 2,513 tonnes/2,473.3 tons (surfaced); 3,540 tonnes/3,484.1 tons (submerged)
**Length:** 117.5m/385ft 6in
**Beam:** 9.44m/31ft
**Armament:** 8 x 533mm/21in tubes (forward), 20 x torpedoes; 1 x 14cm/5.5in (50 calibre) deck gun
**Propulsion:** 2 x diesel engines, 9,246.7kW/12,400hp; 1,491.4kW/2,000hp electric
**Speed:** 23.5 knots (surfaced); 8 knots (dived)
**Complement:** 102 men

LEFT: *I3*, one of the J1 type built in the 1920s, was already tired by the time of their entry into the war, although the class had been fitted with German-supplied diesel engines which provided a higher degree of reliability but no great speed. Consequently, after early battles, they were converted into supply boats, but none survived the conflict. ABOVE: A profile of the J3 boat. Altogether more reliable, these were the largest submarines built by the Japanese prior to the war. Two boats, *I7* and *I8*, were intended as squadron flagships.

# Type J

The J boats, produced through three classes – J1, J2 and J3 – had a heritage dating back to *U139*, and subsequently the prototype of the KD classes, with one big difference – they were much larger. Although the first were launched in the mid-1920s, they were still active in World War II, as can be noted by the naming/numbering system adopted in the same style as U-boats, and the J types were all early numbers. They also represented an early example of the Japanese penchant for monsters of *junsen*, or cruiser, boats that could turn their hand to a number of disciplines.

Even the earliest of the boats had substantial endurance of around 65–70 days and around 40,234km/25,000 miles, although they were inevitably very slow when dived. The diesel engines came from Germany, as did a number of technicians and scientists hired by the Japanese Navy to bring their World War I expertise into their new breed of submarines.

Although deployed in a combat role at the start of World War II, the older boats were converted to cargo vessels and were especially used in the latter stages to ferry supplies to starving Japanese troops in far-flung theatres. Their guns were removed to provide accommodation for a barge used to shift their load to the hungry multitudes ashore.

The earliest of the boats, *I1*, became one such carrier but went into history for a totally different reason. While on a supply run to Guadalcanal, she was set upon by a pair of New Zealand frigates and sank less than a mile offshore. The escaping crew managed to scramble to safety with some of the boat's logs and codes, but left behind the past and future codes. However, the boat did not go under. Her bow was protruding above the water, and repeated attempts by Japanese ships to blow her out of the water failed. The Americans arrived on the scene and managed to salvage a vast hoard of codes and charts, amounting to over 150,000 pages. It provided the intelligence for the American Navy to carry out a series of raids that sent more Japanese submarines to the bottom.

Only two of the last class, J3, were built, utilizing more of the design features of the latest KD types during construction in 1938. They were also the largest submarines completed by Japan before the war, and intended as squadron flagships.

Both carried a seaplane, which were known to have completed intelligence sorties over Pearl Harbor 11 days after the Japanese attack. The Americans sank the two submarines in 1944 and 1945 and indeed none of the J class survived the war.

LEFT: A ceremonial line-up: the J3-class boat *I8* enters German-occupied Brest harbour in 1943. She was later converted to carry the Kaiten human torpedoes, and was sunk off Okinawa in March 1945 by two US destroyers.

## J3 class

**Displacement:** 2,484 tonnes/2,444.8 tons (surfaced); 3,526 tonnes/3470.3 tons (submerged)
**Length:** 109m/357ft 7in
**Beam:** 9m/29ft 6in
**Armament:** 6 x 533mm/21in tubes (bow), 21 x torpedoes; 1 x 14cm/5.5in (50 calibre) deck gun; 1 x seaplane
**Propulsion:** 2 x diesel engines, 8,351.8kW/11,200hp; 2,088kW/2,800hp electric
**Speed:** 23 knots (surfaced); 8 knots (dived)
**Complement:** 102 men

# Type Sen Taka

The Sen Taka I200-class submarine took its name from an abbreviation of the Japanese words for "submarine" and "fast", and it certainly lived up to that christening. It was the only World War II submarine that was easily a match for the German U-boat XXI and was superior to it in the three key areas of power, speed and weaponry.

The Sen Taka stood out among the Japanese boats, which had something of a reputation for slow manoeuvrability and diving. The two 2,051kW/2,750hp engines and streamlined welded hulls provided around 17 knots on the surface, but even more impressive was the coupling with heavy-duty battery cells supplying an impressive 3,728kW/5,000hp electric motor capable of achieving 20 knots – double the speed achieved by contemporary American designs. They were equipped with a snorkel, which allowed for underwater diesel operation while recharging batteries.

Eight boats were laid down, but only three were completed before the end of the war, and the commissioning came too late to see any operational activity. The submarines were designed for mass production, which is why the incoming Americans made sure they kept the Sen Taka secrets to themselves. They took possession of two of the boats and were among the convoy of four Japanese submarines, including two Sen Tokus, which were sailed to Hawaii for inspection by American engineers and designers. Once the inspection was complete, the boats were taken out to deep water near Oahu, torpedoed and sunk by the US submarine *Cabezon* on May 31, 1946, thwarting demands by the Russians to be allowed to examine the boats.

ABOVE: **One of the Sen Taka Sho type submarines, unusually small for the Japanese and designed for fast reaction in the defence of coastal regions. Two dozen boats of this type were planned, the later models with snorkels, but only ten were completed before the war's end, and none became operational.**

BELOW: **A profile of the medium-sized Sen Taka I201 class, the only World War II boat comparable with the German ground-breaking Type XXI, with a streamlined double hull. She was faster submerged than on the surface, and had a snorkel fitted that allowed underwater recharging. Only three were built before Japan's surrender, and none saw service.**

**Sen Taka class**

**Displacement:** 1,270 tonnes/1,249.9 tons (surfaced); 1,427 tonnes/1,404.5 tons (submerged)
**Length:** 79m/259ft 2In
**Beam:** 5.8m/19ft
**Armament:** 4 x 533mm/21in tubes (forward), 10 x torpedoes; 2 x 25mm/0.98in machine-guns
**Propulsion:** 2 x diesel engines, 2,050.7kW/2,750hp; 3,728.5kW/5,000hp electric
**Speed:** 16 knots (surfaced); 19 knots (dived)
**Complement:** 31 men

# Type Sen Toku

LEFT: **A giant, *I402* was one of three Sen Toku boats that were more than 50 per cent larger than the biggest American submarine of that era and with almost twice the range. The three mono-winged aircraft carried by these boats could carry one aerial torpedo or a bomb.** ABOVE: **The open door of the Sen Toku aircraft pod.**

These huge submarines are really a postscript to World War II in that only three of a projected 16 were built before the war ended, and none saw any action. However, they must be noted for their sheer size, power and potential. At the time, they were the largest submarines ever built anywhere in the world, and more than half as big again as the largest American boat of that era, USS *Argonaut*. They were equipped with a snorkel and the latest radar and also had an incredible range of 59,546km/37,000 miles, almost twice that of the American giant, and were specifically designed to bomb the Panama Canal and west coast American cities. However, there were other major differences between the American and Japanese boats, mainly in the hidden added extra that the Japanese had included that, given any further extension of the war, might well have had dire consequences.

As well as a formidable torpedo load, each of the submarines, numbered *I400*, *I401* and *I402*, was equipped to carry three very efficient mono-winged bombers. They were the collapsible Aichi M6A1 Seiran floatplanes that could be stowed in a 36.6m/120ft hangar that would be opened forward to link to the catapult. They could be unpacked, armed and prepared for launch in 45 minutes. The aircraft payload was either one aerial torpedo or an 800kg/1,764lb bomb and claimed a range of up to 1,158km/625 nautical miles and a top speed of 475kph/295mph. The submarine carried enough stock for up to 20 flying missions. Meanwhile, the submarines carried massive on-board anti-aircraft protection with 11 25mm/0.98in anti-aircraft cannon.

For all that, the Sen Toku came too late. They did not come into service until July 1945, and at the end of the month, the first two – *I400* and *I401* – set out on a mission to launch their aircraft in a joint attack with submarines carrying the Kaiten human torpedoes against an American naval anchorage off Ulithi, scheduled for the third week of August. Hostilities ended before they reached the target area, and the attack force returned to base to hoist the white flag.

Two of the boats were subsequently taken to the United States for a detailed examination and were subsequently scuttled in the Pacific a year later.

RIGHT: **Two Sen Toku boats, *I400* and *I401*, were en route to launch their aircraft in kamikaze attacks on the American fleet anchorage at Ulithi on July 26, 1945, when hostilities ceased.**

## Sen Toku class

**Displacement:** 5,140 tonnes/5,058.8 tons (surfaced); 6,456 tonnes/6354 tons (submerged)
**Length:** 121.92m/400ft
**Beam:** 11.9m/39ft
**Armament:** 8 x 533mm/21in tubes (forward), 20 x torpedoes; 1 x 14cm/5.5in (50 calibre) deck gun
**Propulsion:** 4 x diesel engines, 6,338.4kW/8,500hp; 1,491.4kW/2,000hp electric
**Speed:** 18.5 knots (surfaced); 6.5 knots (dived)
**Complement:** 145 men

LEFT: **The great *Surcouf* was a one-off – the world's largest boat – and her mysterious sinking in World War II gave her an unsurpassed record for the largest loss of life in a single submarine accident.** BELOW: **A view of the boat in all her glory, with the twin gun turret – a powerful design feature.**

# Surcouf

ABOVE: **The MB 410 and MB 411 were observation aircraft with a single central float and two small stabilizing floats. They were designed to be carried by *Surcouf* and were easily disassembled for stowage.**

The single boat known as *Surcouf* was literally in a class of her own. Construction dated to 1929, when the French Navy won approval to build what was then the largest and most costly submarine the world had ever seen – and one to which the Japanese would soon pay homage by capping that title. *Surcouf* was commissioned in 1934, there to begin a controversial career and wartime escapades that might have been a plot for a James Bond thriller. She was designed and built as an underwater cruiser to fulfil France's avowed intention of keeping her coastline well defended.

For that mission, the boat carried a hangar for a floatplane, designed primarily for observation. The idea was that the submarine would cruise the coastline unseen, and in the event of being called into action, the giant would rise through the waves and become a warship. She also had the ability to blast the opposition out of the water with no fewer than ten torpedo tubes, and the floatplane was theoretically capable of directing fire for the boat's big guns, up to a range of 24km/15 miles.

There was also a 4.9m/16ft motorboat, and a cargo compartment below was fitted out to hold prisoners. She had a 90-day endurance, and a range of 18,520km/10,000 nautical miles. When the Germans invaded France in 1940, the submarine was in the process of being refitted at Brest, and was ordered to sail immediately even though she had only one engine in operation. Arrangements were made for her and other boats to cross the English Channel to Portsmouth, where, in due course, her commander was called upon to surrender the boat to the British under the terms of Operation Catapult, launched to seize French ships and submarines to prevent them being used against Britain. He refused, and consequently British forces boarded her at Portsmouth, and placed the submarine under the control of the Free French. A fresh crew was mustered, and she began convoy patrol for a while but was then involved in allegations that she was spying for Vichy France, and the British sent two officers to join the boat to keep watch.

The following year, she was used to carry an admiral of the Free French to Canada, and subsequently took on the role of gunboat with other ships to liberate islands off the coast of Newfoundland, helping to remove the Vichy administration. This caused some difficulty with the United States, which had just concluded a deal with the Vichy Government to maintain neutrality in French possessions in the Western Hemisphere. She sailed again amid rumours she was off to liberate islands in the Caribbean. Others stories suggested she was carrying a cargo of French gold.

However, having called at Bermuda on February 12, 1942, the great *Surcouf* was mysteriously lost with all hands. Two theories were proposed, suggesting that she had either been attacked by American bombers, who mistook her for a U-boat, or that she had been accidentally rammed by an American freighter.

### Surcouf

**Displacement:** 3,198 tonnes/3,147.5 tons (surfaced); 4,235 tonnes/4,168.1 tons (submerged)
**Length:** 110m/360ft 11in
**Beam:** 9.2m/30ft 2in
**Armament:** 6 x 550mm/21.7in tubes, 14 x torpedoes; 4 x 400mm/15.75in tubes, 8 x torpedoes; 2 x 203mm/8in twin turret guns, 2 x 37mm/1.46in anti-aircraft cannon, 2 x 13.2mm/0.52in anti-aircraft machine-guns; 1 x MB 411 float plane; 284.5 tonnes/280 tons cargo capacity
**Propulsion:** 5,667.3kW/7,600hp diesel; 2,982.8kW/4,000hp electric
**Speed:** 18.4 knots (surfaced); 10 knots (dived)
**Complement:** 118 men

LEFT: **French submarine *Glorieux* from Series II of the Redoutable class, whose boats were launched between 1928 and 1937, was among the French boats that escaped the Nazi occupation at Toulon on November 27, 1942. She reached Oran to join the Allied cause, surviving the rest of the war before being sold for scrap in 1952.**
BELOW: ***Sfax*, on the other hand, remained faithful to Vichy France but was accidentally torpedoed and sunk by *U37* off Cape Juby, Morocco, in December 1940.**

# Redoutable class

The French produced some extremely attractive submarines in the inter-war years, with sleek lines that sat well on the surface. However, the diversity of boats listed in a dozen classes, including what was then the world's largest submarine, made it difficult to read French policy. Indeed, there seemed to be no precise pattern of development, unlike British policy where an orderly progression was evident, in spite of the latter's financial constraints.

This was partially due to the carryover from World War I in which the 55 French submarines that participated quickly became obsolete. The aftermath saw disagreement between senior naval officers, government figures and international allies as to future requirements. Those who held sway were of the view that the French should concentrate on an armada of surface ships to protect the coastlines, and, where necessary, deliver gunboat diplomacy to their overseas interests in the Far East, Asia and Africa. In consequence, almost three-quarters of French submarines operational at the outbreak of World War II were of designs dating back to the 1920s. Numerically largest of all the classes – and the designated workhorse – was the Redoutable class, whose leader of that name was a 1,422 tonne/1,400 ton submarine known to be a slow diver in emergency situations.

In the event, the scenario confronting French submarines at the onset of World War II and during the early months of the conflict was like no other. There is hardly a story among the nation's 77 boats that prepared for the outbreak of World War II that can be told in the straightforward manner of service, action and fate, from which may be deduced the performance of the boats. Consequently, unlike in World War I when French submarines joined Britain and Australia in the Dardanelles, their impact on the second war bore no comparison.

It was an unholy mess that resulted in great bitterness and the eventual loss of almost 50 per cent of the French submarine fleet, though not necessarily through operational activity at sea. Some boats fell victim to U-boats prior to the French surrender. Next, Winston Churchill announced that the British would sink or impound the French fleet unless the Navy put them out of commission, and eventually carried out that threat against those who did not conform. In a further development in November 1942, fleet commanders at the French naval base of Toulon scuttled their boats to avoid them falling into British or German hands. In the Redoutable class

LEFT: ***Casabianca*, a heroic boat that escaped Nazi clutches at Toulon on November 27, 1942, joined the Allies at Algiers, and there performed sterling work in various intelligence roles, especially in delivering agents into hostile territory.**

BELOW: **Vichy submarine *Poncelet*, scuttled by her crew in December 1940. HMS *Milford* was blockading Gabon and forced *Poncelet* to the surface with depth charges and then shot off the boat's conning tower. The crew was saved.**

BELOW: **The profile of *Casabianca*, which was a standard *Redoutable*-class design.**

of 17 boats, the situation was as follows: eight were scuttled, six were sunk by British destroyers in various locations, one on foreign service was arrested in the Far East and two rejoined the Allies.

There was a famous and mysterious exception to this tragedy. *Casabianca* under the command of Capitaine de Frégate Jean L'Hérminier was among those who took his boat towards the Allies in North Africa.

She was one of the third series of the Redoutable class of boats built and lost under individual type names such as Redoutable, Agosta and L'Espoire dating from 1924. Of the final overall total for the Redoutable series of 31 boats, two were lost before the war and another 23 sunk or scuttled. *Casabianca* had one of the most notable operational records of any French submarine and was the only third series boat to survive the war, operating for the Free French under General de Gaulle's overall command. Commander L'Hérminier risked constant attack as his boat was employed on secret missions delivering agents and saboteurs into Nazi-held Europe and particularly in running supplies and weapons to Corsica for the French resistance. At the same time, he was fully operational as an attack submarine and achieved a number of successes against German shipping. *Casabianca* thus played a key role in the eventual capture of Corsica by the Allies.

RIGHT: ***Conquerant*, a Series II Redoutable of the Agosta type, was launched in 1935 and sunk in November 1942 off Dakhla, Morocco, by two American Catalina aircraft.**

## Redoutable class

**Displacement:** 1,360 tonnes/1,338.5 tons (surfaced); 2,047 tonnes/2,014.7 tons (submerged)
**Length:** 92m/301ft 10in
**Beam:** 8.75m/28ft 8in
**Armament:** 9 x 550mm/21.7in bow tubes; 2 x 400mm/15.7in stern tubes; 1 x 99mm/3.9in deck gun; 1 x 37mm/1.46in anti-aircraft gun; 1 x 13mm/0.51in anti-aircraft gun
**Propulsion:** 4,474kW/6,000hp diesel; 1,491.4kW/2,000hp electric
**Speed:** 17 knots (surfaced); 10 knots (dived)
**Complement:** 63 men

LEFT: **The *Dekabrist* was produced by post-Tsarist Russia, but the shipbuilders of the Soviet Russia had to battle with out-dated technology.**

# Dekabrist class

Russia had fully embraced the concept of submarine warfare prior to World War I when the Tsar's Navy acquired almost 50 boats, including five American Hollands, eight from Simon Lake and others from Britain, Germany and France. They were based around the Baltic and the Black Sea as a defensive measure, but given the political turmoil within the country, her submarines had little impact on events of that era.

Towards the end of the 1920s, however, the new regime of the USSR once again sought external help in establishing a new submarine fleet, employing many German designers and engineers in their new production programme. The result was the Dekabrist class, which represented the new Soviet Union's first attempt at fully home-grown submarines.

The boats, said to be wholly Soviet creations, went into production in 1927 and were overseen by the Bureau of Submarine design, headed by engineer Boris Malinin, who designed the Bars class in 1916–17. They were the first Russian boats with watertight bulkheads and were designed for long-range attacks on enemy communications. They were constructed with riveted hulls of high-strength steel, with a 45mm/1.77in AA gun aft of the conning tower.

In fact, the Dekabrist class proved to be very useful boats for Soviet purposes – manoeuvrable and quick-diving – but eventually falling foul of poor maintenance routines in post-Tsarist Russia and subsequently in constant need of repair. Furthermore, the technology was a good 10 years behind that of the West, and the boats required extensive modernization as World War II approached, when, in time, they would see service in the Baltic and Black Sea. Later in the war, the British loaned some technology, providing an ASDIC-129 sonar and a new torpedo launching system for one of the boats.

The Dekabrist class led the Soviet Union into serious submarine construction, and they next produced the Leninec class of six minelaying boats, laid down in 1928 and 1930 based on the same hull. In this class, which entered service in 1933, the two stern torpedo tubes were discarded in favour of tubes for 20 mines.

### *Dekabrist*

**Displacement:** 910 tonnes/895.6 tons (surfaced); 1,190 tonnes/1,171.2 tons (submerged)
**Length:** 76.5m/251ft
**Beam:** 6.3m/20ft 8in
**Armament:** 8 x 533mm/21in tubes (6 x bow, 2 x stern), 5 reloads; 1 x 45mm/1.77in anti-aircraft gun, 500 rounds
**Propulsion:** 820.3kW/1,100hp diesel; 391.5kW/525hp electric
**Speed:** 15.5 knots (surfaced); 8 knots (dived)
**Complement:** 53 men

LEFT: **Mini-submarines had been left behind by the major seafaring nations. Not so with the Soviets, who produced this small craft in vast numbers.**

# Malutka class

The class name Malutka, meaning "baby", was an apt description. Almost 100 were eventually built, and they were designed primarily as coastal submarines on a mass-production basis, intended largely to defend naval bases or, in times of war, to blockade enemy shipping routes and harbours. However, for the USSR, the Malutkas had a distinct advantage in that they were also fully transportable by rail, thus enabling them to be carried towards any theatre of war across the vast Russian terrain to the nearest sea to begin their patrols – or, in some cases, the nearest lake. A number of the boats were said to have been transported to Lake Ladogo, near Leningrad, for operations. The M class was built in four versions, each following on from the last with improvements, modifications and lengthening.

### Malukta class

**Displacement:** 153 tonnes/150.6 tons (surfaced); 187 tonnes/184 tons (submerged)
**Length:** 37.10m/121ft 9in
**Beam:** 4.15m/13ft 7in
**Armament:** 2 x 457mm/18in tubes, 2 x torpedoes (4 x tubes, 4 x torpedoes on later versions)
**Propulsion:** 507.1kW/680hp petrol; 186.4kW/250hp electric
**Speed:** 13 knots (surfaced); 7.4 knots (dived)
**Complement:** 18 men (30 for the XV version)

LEFT: **The Scuka class was a candidate for mass production in Soviet Russia. By the time of World War II the Soviets had a large number to call upon, but the submarines were nowhere near as efficient as the German U-boats and consequently suffered the consequences.**

## Scuka class

**Displacement:** 558 tonnes/549.2 tons (surfaced); 689 tonnes/678.1 tons (submerged)
**Length:** 57.50m/188ft 8in
**Beam:** 7.15m/23ft 6in
**Armament:** 6 x 533mm/21in tubes (4 x bow, 2 x stern), 12 x torpedoes; 2 x 45mm/1.77in deck guns
**Propulsion:** 894.8kW/1,200hp petrol; 335.6kW/450hp electric
**Speed:** 12.5 knots (surfaced); 6.5 knots (dived)
**Complement:** 50 men

# Scuka class

This medium-sized submarine became one of the main workhorses of the Soviet fleet during World War II, passing through three versions, with 91 eventually operational. When originally planned, these boats represented the first major USSR programme of attack submarines designed to be "positioned against closed and defended enemy theatres of war" and, in due course, would be assigned for operations with four principal Soviet Navy fleets in the North, the Pacific, the Black Sea and the Baltic. It was in the latter two stations that they were eventually to suffer the heaviest casualties against the Nazis, where almost 70 per cent of their front-line boats were lost. World War II operational records also include confrontations with the Japanese in the Pacific, where their performance was outstanding.

LEFT. **Ironically emanating from German design work and Dutch engineering, the Soviets began producing what they proudly called the Stalinec class from around 1934 and into the war years. Some boats were still operative in the 1950s.**

## Stalinec class

**Displacement:** 829 tonnes/815.9 tons (surfaced); 988 tonnes/972.4 tons (submerged)
**Length:** 77.80m/225ft 3in
**Beam:** 9.85m/32ft 4in
**Armament:** 6 x 533mm/21in tubes (4 x bow, 2 x stern), 12 x torpedoes; 2 x 45mm/1.77in deck guns
**Propulsion:** 894.8kW/1,200hp petrol; 335.6kW/450hp electric
**Speed:** 12.5 knots (surfaced); 6.5 knots (dived)
**Complement:** 50 men

# Stalinec class

This boat was one that emerged in the 1930s as a result of direct co-operation between the USSR and Germany – a partnership that the latter nation subsequently came to regret. The boat emerged from a prototype originally built at Karhatena by a Dutch subsidiary using German designers for the Spanish Navy. When the Spanish subsequently cancelled the order, the Soviet Navy was invited to inspect the boat for possible inclusion in their fleet. An agreement was reached whereby plans and drawings for the boat would be purchased by the USSR. Later, a party of Soviet designers and engineers travelled to Bremen and Karhatena to discuss with the German engineers design changes to accommodate Soviet requirements. Thus, the S class was born, utilizing Soviet parts and machinery, built in five Soviet shipyards.

# 600 class

The build-up of Italian submarines began in earnest in 1925 under Mussolini and included a number of medium to large boats whose quality may well have been hampered by financial constraints in the aftermath of World War I. However, in the early 1930s, the Italians began a substantial programme of construction following an international disarmament conference in London in 1930 that set limits on the size – but not quantity – of boats built by the naval powers.

The conference was called amid international concern about the growth in submarine fleets, which had arisen since a 1921 conference that placed restrictions on the expansion of naval fleets but excluded submarines. Britain, the United States, Japan and France had already been demonstrating to the world some grandiose projects for international operations in the late 1920s. Thus, in 1930, defining limits were set for coastal and oceanic boats, 610 tonnes/600 tons for the former and 2,032 tonnes/2,000 tons for the latter. The Italian Navy took a somewhat pragmatic approach to the new regulations and responded by introducing what became known as the 600 class. This, in effect, set them on course to build a large number of boats of the coastal classification, which were well suited for their principal area of operations in the Mediterranean, under the generic name of the 600 class.

They were introduced progressively in five types, built to a master drawing with some deviations for each under the names Argonauta (after the first boat of the series), Sirena, Perla, Adua and Platino. The result was that by the time of Italy's entry into World War II, Mussolini was able to deploy a diverse range of modern submarines made up of 50 large, 89 medium, two cargo and assorted midget boats, to which further additions came during the war years.

The 600 class, built with remarkable speed by various shipyards across Italy, was at the heart of the nation's submarine force, and undoubtedly the most successful. Furthermore, the stream of boats emerging through this period was competitive and well managed. Unfortunately, the lead member of the group was not among those who participated. When Mussolini declared war on Britain on June 9, 1940, the *Argonauta*, under the command of Lt Vittorino Cavicchia Scalamonti, was one of the 55 boats patrolling the Mediterranean.

Her station was to cover the British naval base of Alexandria and Suez where the Royal Navy was already engaged in anti-submarine activity. Eleven days into the patrol, she took a heavy bombardment of depth charges, and the damage included her periscope. She was forced to return to the Italian base at Tobruk, but there were no facilities for such repairs, and she was sent on to Taranto. Thereafter, nothing more was heard of her, and she was presumed lost with all hands.

ABOVE: ***Argonauta*, one of the most successful Italian submarines, led the way in the pre-war 600 class, which consisted of a series of five types produced progressively under the names of Argonauta, Sirena, Perla, Adua and Platino. In all, 59 submarines were built by various shipyards, and these differed in their minor detail.**

LEFT: ***Nichelio*, one of the Platino types in the 600 series. The 600-class boats were of a very decent and sleek design that evolved and improved through 59 units.**

## Argonauta Type

**Displacement:** 656 tonnes/645.6 tons (surfaced); 797 tonnes/784.4 tons (submerged)
**Length:** 61.5m/201ft 9in
**Beam:** 5.65m/18ft 6in
**Armament:** 6 x 533mm/21in tubes; 1 x 102mm/4in deck gun; 2 x 17.2mm/0.68in anti-aircraft guns
**Propulsion:** 932.1kW/1,250hp diesel; 596.6kW/800hp electric
**Speed:** 14.5 knots (surfaced); 8.2 knots (dived)
**Complement:** 36 men

ABOVE: **The *Marcello* was the first one of a series of 11 boats of the successful Italian class of the same name, from which the Marconi design emanated. The six submarines of the Marconi class became renowned for "good seaworthiness and easiness of deployment, especially after the reduction in size of the conning tower displayed on the *Marcello*".**

# Marconi class

Among the new classes introduced by the Italian Navy in the 1935–40 period, Marconi boats had a good record in seaworthiness and ease of deployment, although later alterations when they were moved to Bordeaux in support of U-boat operations included enlargements that affected their buoyancy to what some naval specialists regarded as dangerously low levels. Even so, Regia Marina Italiana today maintains that they were "surely among the best vessels produced by Italy". One them, the *Leonardo Da Vinci*, was credited with the highest number of sinkings among Italian submarines, second only to U-boats. Infamously among them (from both points of view) was the sinking on March 14, 1943, of the transatlantic liner *Empress of Canada* with 3,000 British soldiers and 500 Italian prisoners of war aboard. The submarine managed to save only one, a doctor named Vittorio Del Vecchio.

The six submarines of the Marconi class operated in the Atlantic and Indian oceans for the majority of the war, sinking an aggregate total of 38 ships with a total of 219,697 tonnes/216,227 tons and damaging another 17 for a total of 118,558 tonnes/116,686 tons. None of them survived the war, five having been lost with all hands in the Atlantic. The sixth, *Torelli*, had an interesting journey towards her final demise. She was captured by the Japanese in Singapore after Italy's surrender in 1943 and was subsequently handed over to Germany's U-boat command, who renamed her *U(IT)25*. The boat operated with a German/Italian crew until Germany's own surrender, when she was hastily recovered by Japan, again operating with a mixed crew, including some Italians. The Americans subsequently sank her off the Japanese coast near Kobe.

ABOVE: **A profile of the *Leonardo da Vinci* from the Marconi class, credited with the highest number of sinkings among Italian submarines. This sink rate was comparable to U-boat performance and was a higher rate than most of the Allied submarine forces.**

## Marconi class

**Displacement:** 1,172 tonnes/1,153.5 tons (surfaced); 1,465 tonnes/1,441.9 tons (submerged)
**Length:** 70.04m/229ft 9in
**Beam:** 6.82m/22ft 5in
**Armament:** 8 x 533mm/21in tubes; 4 x 13.2mm/0.52in anti-aircraft guns
**Propulsion:** 2 x diesel engines, 2,684.5kW/3,600hp; 2 x electric motors, 1,118.6kW/1,500hp
**Speed:** 18 knots (surfaced); 8 knots (dived)
**Complement:** 57 men

# Index

## Acknowledgements

Research for the images used to illustrate this book was carried out by Ted Nevill of TRH Pictures – Cody Images, which supplied the majority of the pictures. The publisher and Ted Nevill would like to thank all those who contributed to this research and to the supply of pictures: ArtTech; BAE SYSTEMS; Chrysalis Images; Jeremy Collins; DCN; General Dynamics Electric Boat; Howaldtswerke-Deutsche Werft; James Fisher Defence; Kockums; Naval Photographic Club and Dr Duncan Veasey; Norwegian Ministry of Defence; Photographic Section, Naval Historical Center, Washington, DC, USA; Robert Hunt Library; Royal Navy Submarine Museum; David Saw; Still Pictures, National Archives & Records Administration, College Park, Maryland, USA; US Navy.

## Key to flags

For the specification boxes, the flag that was current at the time of the vessel's commissioning and service is shown.

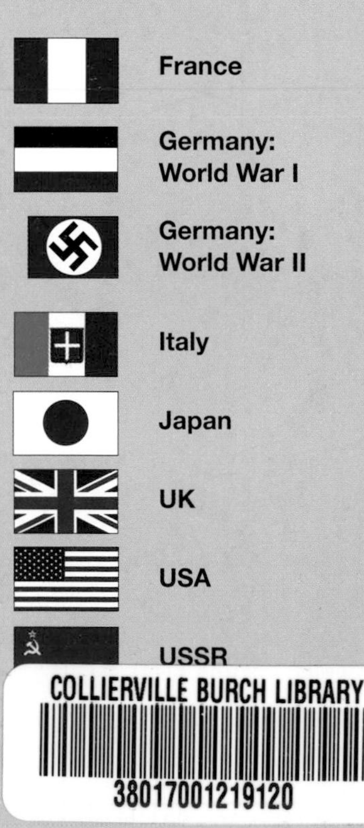